Nicholas M. Hentz, Edward Burgess, James H. Emerton

## The Spiders of the United States

a collection of the arachnological writings of Nicholas Marcellus Hentz

Nicholas M. Hentz, Edward Burgess, James H. Emerton

**The Spiders of the United States**
*a collection of the arachnological writings of Nicholas Marcellus Hentz*

ISBN/EAN: 9783337400842

Printed in Europe, USA, Canada, Australia, Japan

Cover: Foto ©berggeist007 / pixelio.de

More available books at **www.hansebooks.com**

# OCCASIONAL PAPERS

OF THE

# BOSTON SOCIETY OF NATURAL HISTORY.

II

BOSTON:
PRINTED FOR THE SOCIETY.
1875.

# THE SPIDERS OF THE UNITED STATES.

A COLLECTION OF

# THE ARACHNOLOGICAL WRITINGS

OF

## NICHOLAS MARCELLUS HENTZ, M.D.

EDITED BY

### EDWARD BURGESS,

WITH NOTES AND DESCRIPTIONS BY JAMES H. EMERTON.

BOSTON:
BOSTON SOCIETY OF NATURAL HISTORY.
1875.

# PREFACE.

Among the pioneers in the study of American Entomology Nicholas Marcellus Hentz must take a prominent position. That he was an entomologist of general attainments, his correspondence with Harris — already familiar to the readers of the first volume in this series of "Occasional Papers" — bears abundant witness, but with the study of American Arachnology his name and his writings are almost exclusively associated.

In selecting the Spiders for his special study, he found not only an interesting, but an almost entirely unexplored field. Before his time, with the exception of a few accidental descriptions scattered through the works of writers, for the most part European, nothing relating to North American Spiders had been published. This was perhaps, on the whole, fortunate, for as he lived for the greatest part of the time in places where great libraries were inaccessible, the danger of repeating the work of others was avoided. But whether his choice was accidental or predetermined, he began, soon after settling in America, a diligent study of these insects, and devoted all the time

which could be spared from the busy profession of teaching, to
the observation of their habits, and to the collection, description
and representation of the various species.

After publishing a few short papers in Silliman's Journal,
and in the Journal of the Philadelphia Academy of Arts and
Sciences, he brought together his extensive series of notes and
paintings and offered them to the Boston Society of Natural
History for publication in its Journal.

The Society readily undertook the publication of such a valuable contribution to American Natural History, but the cost
of illustrating such a number of forms rendered it necessary to
extend its publication over a long series of years. Thus, as I
understand from his son, Hentz never revised or even saw but
the first one or two parts of his work. In the course of publication a number of drawings, illustrating details of structure for
the most part, as well as a few notes, were set aside, probably
to lessen, as far as possible, the expense. This material was,
however, later collected by the former Secretary of the Society,
Mr. S. H. Scudder, and published, in the form of a supplement
to Hentz's monograph, in its 'Proceedings.'

As many parts of this work have been long out of print,
the Council of the Society has determined to republish in a
connected form all of Hentz's arachnological writings, and thus
to prepare the foundation necessary for future work in a field in
which as little progress since, as before, Hentz's time has been
made. To render the work as valuable as possible, Mr. J. H.
Emerton, who has paid much attention to the study of our
native spiders, has added a very considerable number of notes,
descriptions and synonymical remarks, which will prove of great
assistance to the student. Two new plates from his skillful

pencil farther illustrate the subject. These notes have been printed in smaller type and, with Mr. Emerton's initials, inclosed in brackets. The Society is also indebted to Mr. Wm. E. Holden, of Marietta, Ohio, for some additional notes to which his name is appended.

The proportion of species identified by Mr. Emerton is, perhaps, small, a fact not to be wondered at when we consider how many Southern forms were described by Hentz, for the recognition of which, extensive collecting in the South is necessary. Among the species identified, a number have proved identical with European forms, and Hentz's names must yield to those longer established.

To aid the student in referring to the original place of publication, the pagination of the latter has been inserted in black-faced type in the text of the present work. For the same reason the original numbers remain on the plates, although these are now referred to by new numbers for the sake of convenience. As the stones from which the lithographic plates were taken were destroyed, and as unfortunately several of the copper plates are also missing, it has been necessary to reproduce nearly half of the plates in this edition by some method of photography, a work which has been entrusted to Mr. E. Bierstadt, of New York. Although the Alberttype plates by no means equal the originals either in beauty or in clearness, it is hoped that most of the figures will be recognizable without difficulty.

After Hentz's death his collection of spiders came into the hands of the Society, but has long since been almost entirely destroyed. The remains now consist of portions of sixty specimens gummed on cards, and of these, only twenty-seven can be identified by their labels. In the absence therefore of the type specimens the beautiful collection of Hentz's paintings and drawings, carefully preserved in the Society's library, is the surest basis remaining for the identification of his species. So far as it has been possible to compare those drawings with specimens they are correct in colors and markings, but seem to have been drawn without measurements, and the legs in nearly all cases are too short. The figures of the eyes are generally good, but those of the maxillæ and mandibles are of little use.

Having thus explained the origin, purpose and plan of this volume, it remains only to present a sketch of Prof. Hentz's life, which, brief as it is, will be, I hope, interesting to Entomologists.

The materials enabling me to do this have been kindly furnished by his eldest son, Dr. Charles Arnould Hentz, of Florence, Alabama.

NICHOLAS MARCELLUS HENTZ was born in Versailles, July 25, 1797. His father, an advocate by profession, was actively engaged as a politician at the time of Hentz's birth, and had been, shortly before this event, obliged to flee from his home in Paris, and to conceal himself in Versailles under the assumed name of Arnould. To the agonizing fears and alarms which his mother was obliged to undergo during this period, Hentz was wont to attribute the peculiarities of his nervous system, which were, as will be seen, very remarkable.

At the early age of between twelve and fourteen years he began the study of miniature painting, for which he showed great talent and became highly proficient. He soon, however, became interested in medicine and entered the Hospital Val-de-

Grâce as a student. His son still possesses, in an old parch-
ment-covered memorandum book, the following record in
Hentz's then boyish hand-writing, " le vendredi 22 octobre
1813, j'ai été au Val-de-Grâce, M. Hentz. " There he remained, busied with his studies and duties as hospital assist-
ant, until the fall of Napoleon, when his father was proscribed
and obliged to flee to America, whither Nicholas and one of his
brothers accompanied their parents.

The party sailed from Havre-de-Grâce, in the bark " Eu-
gene," Jan. 22, 1816, and arrived in New York City on March
19. Here and in Elizabeth Town they spent a few weeks in
collecting their personal effects and making arrangements to
move into the interior, an undertaking which was then quite
formidable. They arrived in Wilkesburg, Pennsylvania, in the
latter part of April, where it is probable that Hentz's parents
finally settled.

Hentz himself for several succeeding years lived in Boston
and Philadelphia, where he taught French and miniature painting. He also passed a short time on Sullivan's Island, near
Charleston, S. C., as tutor in the family of a wealthy planter, a
Mr. Marshall. All this time, whenever leisure hours allowed
it, he was engaged in entomological studies, directing his special attention, as has already been said, to the spiders. While in
Philadelphia he became intimate with the naturalist, Le Sueur.
Le Sueur was accustomed to etch his own drawings, and having
the use of his press, etc., Hentz made etchings of some of his
spiders, as well as of an alligator, which he had dissected to
study the nature of its circulation.

In the winter of 1820-21, he attended a course of medical
lectures in Harvard University, but finally abandoned the study

of medicine, and engaged himself as teacher in a school for boys at Round Hill, Northampton, Mass., where Bancroft, the historian, was also employed. Here he was married to Miss Caroline Lee Whiting, the daughter of Gen. John Whiting, of Lancaster, on Sept. 30, 1824, and Mrs. Caroline Lee Hentz became afterwards well known as a poet and novelist.

Soon after their marriage, Hentz and his wife removed to Chapel Hill, N. C., where he had been offered the chair of modern languages in the State University. In 1830 he moved to Covington, Ky., to take charge of a female seminary, and a year or two after, to Cincinnati, where he was similarly engaged. "A graceful allusion," writes Dr. Hentz, " is made to them during this time, in Mansfield's 'Life of Daniel Drake,' 1855, p. 226."

In 1834 they again removed to Florence, Ala., and there for eight years conducted a flourishing school, the "Locust Hill Female Academy." In 1842 they went to Tuskaloosa, and in 1846 to Tuskegee, both towns in Alabama, and the following year to Columbus, Georgia, all the time engaged in similar teaching.

In the latter place in 1849, Hentz's health began to fail, his whole nervous system giving away. He grew gradually more and more infirm, and became a regular user of morphine, which he took daily for several years before his death. He moved, finally, to the residence of his son Charles in Mariana, Fla., where he died November 4, 1856.

In person, Prof. Hentz was a small, spare man, about five feet and a half in height, and weighing only one hundred and ten or one hundred and fifteen pounds. Although of a genial, affectionate, and generous nature, his peculiarly nervous organization made him often morbidly sensitive and suspicious, and a prey to groundless fears, which not a little marred his enjoyment of life. He was educated in the Roman Catholic religion, but in 1835 joined the Presbyterian Church. During his whole life he had a most remarkable habit of suddenly resorting to mental, ejaculatory prayer. Without regard to circumstances, in any place, or among any people, he would sometimes, without apparent external reason, take off his hat, or perhaps drop on his knees, press his hands to his forehead, and raising his eyes heavenward, remain in more or less protracted prayer. He had also several regular places for this singular custom, as before his study-door, which he never entered without stopping a moment in silent prayer, and beneath a picture he had made of the "All-seeing Eye"; indeed, the constant pressure of his forehead against the wall in these places left an indelible mark.

He was extremely fond of exercise, and his Saturday half-holidays were invariably spent in long walks with his sons in the woods, carefully collecting insects and observing their habits. For amusement he delighted in fishing and gunning.

He was a great friend of Dr. Thaddeus Wm. Harris (one of his sons, Dr. T. W. Hentz of Columbus, Ga., was named after him), and after separated from him by his own removal to the South, a constant correspondence, mostly entomological, was kept up between the two friends, a portion of which, as already remarked, was published in the first volume of this series.

1821. A notice concerning the Spider whose web is used in Medicine (*Tegenaria medicinalis*). Journ. Phil. Acad. Nat. Sci.. II, p. 53–55.

1825. Some observations on the anatomy and physiology of the Alligator of North America. Trans. Am. Phil. Soc., II, pp. 216–228.

1825. Description of some new species of North American Insects. Journ. Phil. Acad. Nat. Sci., V, p. 373–375.

1829. The same paper in Férusac's Bulletin des Sciences Naturelles, xviii, p. 475–476.

1830. Description of eleven new species of North American Insects. Trans. Amer. Phil. Soc., III, p. 253–258.

1830. Remarks on the use of the Maxillæ in Coleopterous insects, with an account of two species of the family Telephoridæ (*Chauliognathus marginatus* and *C. bimaculus*), and of three of the family Mordellidæ (*Rhipiphorus dimidiatus, R. limbatus* and *R. tristis*) which ought to be the type of two distinct genera. *Ibid.* pp. 458–463.

1832. On North American Spiders. Silliman's Journal of Science and Arts, xxi, pp. 99–152.

1833. Enumeration of Spiders of the United States. Hitchcock's Report on the Geology, etc., of Massachusetts, p. 564. [Contains only the list of genera published in the preceding paper.]

1835. List of Spiders of the United States. *Ibid.* Second edition. [This edition enumerates one hundred and twenty-five species, mostly by name, and arranged under the genera given in the first edition. The species are those described in the Journal Bost. Soc. Nat. Hist., and the paper, giving no additional information, is not reprinted here.]

1841. Description of an American Spider (*Spermophora meridionalis*) constituting a new subgenus of the tribe Inæquitelæ, Latreille. *Ibid.* XLI, pp. 115–117.

1841. Species of Mygale of the United States.[1]   Proc. Bost. Soc. Nat. Hist., I, pp. 41-42.

1842. Description and Figures of the Araneides of the United States. Journal Bost. Soc. Nat. Hist., IV, pp. 54-57. Pl. 7. Continuation, pp. 223-231. Pl. 8.

1844. Continuation, *ibid.*, pp. 386-396. Pl. 17-19.

1845. Continuation, *ibid.*, V,[2] pp. 189-202. Pl. 16, 17.

1846. Continuation, *ibid.*, pp. 352-369. Pl. 21, 22.

1847. Continuation, *ibid.*, pp. 444-478. Pl. 23, 24, 30, 31.

1850. Continuation, *ibid.*, VI, pp. 18-35. Pl. 3, 4. Conclusion, pp. 271-295. Pl. 9, 10. ,

1867. Supplement to the Descriptions and Figures of the Araneides of the United States. Edited by S. H. Scudder. Proc. Bost. Soc. Nat. Hist., XI, pp. 103-111, with two plates.[3]

[1] This paper contains merely the descriptive portion of the first part of the series of articles in the Journ. B. S. N. H. (IV, 57). The species described are; p. 41, *Mygale truncata*; p. 42, *solstitialis, carolinensis, gracilis,* and *unicolor.*

[2] The plates were unfortunately wrongly numbered in this volume of the Journal, the numbers 16, 21, 22, 23 and 24, being used twice.

[3] The text of this Supplement has been, for the sake of conciseness, distributed through the present volume, always in brackets and with the word *Supplement* appended. To enable the student to refer to the original place of publication it may be observed that p. 103 contains the notes from *Katadysas* to *Micromata carolinensis;* p. 104, *M. marmorata* to *Attus familiaris;* p. 105, *Attus fasciolatus* to *A. vittatus;* p. 106, *Epiblemum* to *Clubiona pallens;* p. 107, *C. piscatoria* to *Tegenaria;* p. 108, *Agelena* to *Epeira labyrinthea;* p. 109, *E. mutura* to *E. vulgaris;* p. 109, *Phyllyra* to *Theridion pullulum;* p. 110, *Theridion roscidum* to *T. verecundum.* The species are arranged alphabetically under their proper genera.

[From the Am. Jour. Science and Arts, xxi, 99.]

ART. XIII.   ON NORTH AMERICAN SPIDERS.   By N. M. HENTZ, Principal of the Female Academy at Covington, Kentucky, and late Professor of Modern Languages in the University of North Carolina.

### LETTER TO THE EDITOR.

AMHERST COLLEGE, August 22, 1831.

PROFESSOR SILLIMAN:

*Sir* — Some time since I addressed a request to Nicholas M. Hentz, Esq., then Professor of Modern Languages in the University of North Carolina, and now Principal of the Female Seminary in Covington, Kentucky; that he would furnish me with a list of the Araneïdes [100] found in Massachusetts; as I wished, in the execution of a commission from the government, to obtain as complete a zoological catalogue for the State as was practicable. He not only complied with my request, but sent so full a view of North American Spiders, with so many valuable notes, that I immediately requested and obtained permission to send the whole for insertion in your Journal. If your views of the value of the paper correspond with my own, I shall hope you will give it a place in the next number.                With much respect,

EDWARD HITCHCOCK.

ARANEÏDES, (Latreille). *Aranea*, (Linnæus).

| | | No. of species. |
|---|---|---|
| **TETRAPNEUMONES** 8 eyes; 4 mammulæ, 2 very short; tooth of the mandibulæ (chélicères) articulated downward, | Oletera | 2 |
| 6 or 8 eyes; 6 mammulæ; tooth of the mandibulæ articulated laterally, | Filistata? Dysdera Segestria? | 1 1 1 |
| **DIPNEUMONES** — *Araneïdes spinning webs, or wandering; never with all the following characters united, 4th pair of legs longest, eyes in two rows both bent upward, and six mammulæ of which two very long.* — Araneïdes forming no silken habitation, wandering; legs, 4th pair longest; eyes 8, in two rows, never both bent downward; 6 mammulæ, 2 very long, | Herpyllus | 8 |
| Araneïdes making a web, sedentary, | Clubiona Tegenaria Agelena Theridium Pholcus Linyphia Tetragnatha Epeïra Mimetus | 6 2 2 5 1 5 2 26 1 |
| Araneïdes making no web for a constant residence, | Thomisus Sphasus Dolomedes Lycosa Attus Epiblemum | 8 3 6 11 29 2 |
| | Species not included in Attus | 3 |
| | | 125 |

OLETERA, (Walckenaer). *Atypus*, (Latreille).

Eyes 8, Pl. 2, fig. 1 *b*; 4th pair of legs longest. Two species. A male was found at Round Hill, which is totally black. A female of another [101] species was found in North Carolina. It is of a glossy brown, except the abdomen, which is piceous. The palpi are so elongated as to have all the appearance of legs.

## Filistata ? (Latr.).

Eyes 8, nearly equal in size, Pl. 2, fig. 6 *b*; legs 1. 4. 2. 3. lingua surrounded by the maxillæ, which are bent and pointed at their apex. The mandibulæ (chelicercs) are united together, so as to have no reciprocal motion, except by means of their teeth, which are very short. This is a remarkable character, which induces me to believe this may not belong to Filistata, and in that case must be the type of a new genus. These spiders form white silk tubes, in walls and crevices of rocks; the orifices of those tubes are spread and closely fixed on the edges of the stones which make their abodes. I kept several alive under glass, and witnessed the reproduction of their legs. The part torn off does not grow gradually; but when the spider casts its skin, that part comes out with all its joints from the skin, only somewhat shorter than it was before. It is important to observe that, owing to this fact, the character derived from the respective lengths of the legs is often deceptive, as spiders in their conflicts often lose their legs, and frequently offer the characters of two different genera on that account. It is therefore necessary to compare many specimens and the two sides of the spider; but that excellent character ought not to be given up. One species.

## Dysdera, (Latr. Walck.).

Eyes 6, Pl. 2, fig. 2 *b*; legs 1. 2̅. 4̅. 3. lingua truncated. One species. A male and a female were sent to me by Dr. T. W. Harris, one of the most accurate and indefatigable entomologists in this country, who found them in a cavity under ground.

## Segestria ? (Latr.).

Eyes 6; legs 1. 2. 4. 3. lingua longer than broad; maxillæ elongated, narrower above. One species, which is found under the bark of trees in silk tubes. I marked this with a point of interrogation, because Latreille in his last work [1] excludes this from his [102] Tetrapneumones. But the affinity to Dysdera is such that I think they ought not to be separated. And the next genus which I have established seems to me to be the link that unites the preceding to the Dipneumones and goes better after Segestria than before, in a natural arrangement.

## HERPYLLUS, (Mihi).

Eyes 8, Pl. 11, figs. 2, 15; two rows, one or both curved upward; legs 4. 1. 2. 3. rather stout and short; lingua large, short, nearly triangular or slightly truncated; maxillæ straight, wider near the apex, not sensibly serrated, tooth moderately long; cephalothorax ellipsoid, gradually narrowed before, abdomen nearly of the same form. Making no web or tube for their dwellings, but wandering for prey, and running with great velocity. Eight species. I have never found their cocoon. The great affinity between this genus and Tegenaria and some Clubionæ requires that it should be placed here. This *may* belong to the Diplotoxops of Mr. Rafinesque; but as he makes the first pair of legs longest, and as his generic description is vague and incorrect in many respects, for instance, in its having a character derived from the palpi which he may not know is a mere sexual distinction, I could not and ought not adopt his name. Several species are common in the United States, particularly a small black one, found under stones in highways; and a blackish one with a white band on the cephalothorax, a band on the abdomen, beginning at base and reaching the middle, and a spot near the apex white. This one attains a great size, and is found in houses, under stones, planks, the bark of decaying trees, etc. I call it *H. ecclesiasticus*, and the former *H. ater.*

[1] Familles Naturelles du Règne Animal. Paris, 1825. 1 Vol. 8vo.

## CLUBIONA, (Latr.).

Eyes 8, in two' rows curved variously; legs 1. 4. 2. 3. or 4. 2. 1. 3. or 4. 1. 2. 3. lingua truncated. Araneïdes forming silk tubes in leaves which they twist, or under the bark of trees. Six species. Most species fly about in the air, by means of a long thread, at the end of which they suspend themselves, and which is borne by the wind, sometimes raising them to a great height.

## TEGENARIA, (Walck.). *Aranea*, (Latr.).

Eyes 8, Pl. 11, fig. 31; legs 4. 1. 2. 3. Making in houses, cellars and other dark places, the common webs, which are spread horizontally, [103] and have a tube, usually concealed in a hole or crevice, for the reception of the spider. This is the common house spider, the web of which is narcotic, and has been administered internally in some cases of fever with success. It is also effectual in stopping the blood of cuts and slight wounds. Two species only are known to me.

## AGELENA, (Walck.). *Aranea*, (Latr.).

Eyes 8, Pl. 12, fig. 1; legs 4. 1. 2. 3. Making in the fields webs which are spread horizontally, and at the upper part of which is a tube for the retreat of the spider. Two species. Differs from the preceding only in the arrangement of the eyes, and in its preferring the open air to dark retreats.

## THERIDIUM, (Walck.).

Eyes 8, Pl. 16, figs. 1, 2, 4; legs 1. 4. 2. 3. lingua short; maxillæ elongated, inclined over the lingua. Making a web formed of threads crossed irregularly in every direction. Five species. One of them, *Theridium verecundum* (my catalogue), is entirely glossy black, except two crimson spots under the abdomen, the last of which is sometimes continued on the back in the form of a band. It is common in the Southern States, and is well known, as the people there believe its bite is very poisonous. That spiders are all supplied with a poisonous fluid conveyed in their fangs, there can be no doubt; but I cannot assert that this is more dangerous than another, for persons who do not study Natural History are apt to confound objects of a different nature. A respectable physician, however, pointed out this species to me as the one, and told me that in every instance he could arrest the violent symptoms arising from its bite by inducing a reaction in the system, and frequently had produced instant relief with a glass of brandy. Most species of this genus are the common prey of the several species of Sphex called *dirt-daubers* in the South, on account of their making clay nests, in which they enclose with their progeny from twenty to thirty spiders, which serve as food to the young larvæ.

### Pholcus, (Walck.).

Eyes 8, Pl. 17, fig. 7; legs very long and slender, 1. 2. 4. 3. lingua short, triangular; maxillæ long, inclined over the lingua. Making a loose web. One species. Inhabiting the ceilings of houses. I seldom met with it at a distance from the Atlantic coast. [104]

### Linyphia,. (Latr.).

Eyes 8, Pl. 18, figs. 23, 24; legs 1. 2. 4. 3. lingua short, triangular; maxillæ short, wider above. Making a horizontal web on bushes, with another one surrounding it above and below, constructed of threads crossed in every direction, as that of Theridium. The spider holds itself downward, under the horizontal web. Five species, all small.

### Tetragnatha, (Latr.).

Eyes 8, in two rows nearly parallel; legs very long and slender, 1. 2. 4. 3. lingua short, rounded; maxillæ and mandibulæ very long. Making a spiral web with concentric threads. Two species, inhabiting the vicinity of water. .The form of one of the species is rendered horrible, by the size of its mandibulæ, which are longer than the cephalothorax, and armed with numerous prongs, and with fangs which are nearly as long, so that the jaws nearly equal in length the rest of the body. The males are better and more stoutly armed than the females.

## Epeira, (Walck.).

Eyes 8, Pl. 13, fig. 88 ; legs 1. 2. 4. 3. lingua short, rounded ; maxillæ short, rounded. Making a spiral web with concentric threads. Twenty-six species. These spiders are known to every body. They are seen towards night busily engaged in making their admirably contrived webs, in the middle of which they wait for their prey during the night, but usually take shelter during the day under some leaf or crevice, furnished with a tent made of loose threads. The endless variety of forms and habits of the species of this genus, have given rise to natural subdivisions, which are useful, as the number of species is very great. Many have thorns, tubercles and various projections, which give them a fantastic appearance. The form of their cocoons also varies much. Some attach them to their web in a string.

## Mimetus, (Mihi).

Eyes 8, Pl. 18, fig. 33 ; legs long, slender, 1. 2. 4. 3. lingua short, triangular; maxillæ long, slender, pointed at tip, inclined over the lingua ; mandibulæ very long and slender. Making a double web, like that of Theridium and that of Epeïra connected. The spiral regular web is attached behind by innumerable threads to the irregular one, in the upper part of which a tent is constructed with dried leaves, [105] under which the spider takes shelter in the day time. One species, *M. syllepsi-*

*cus* (my catalogue), of a pale green color varied with black on the cephalothorax and abdomen, tips of the four anterior thighs with a black ring, feet very hairy; inhabiting damp woods. The legs and the eyes correspond with Epeïra, but the trophi, except the mandibulæ, are those of Theridium; and the web and habits participate of both. The long and slender mandibulæ are peculiar to this. The cocoons resemble a plano-convex lens, are of a pale brown color, and are attached in the middle, one above another, in the tent which the spider inhabits. It is evident that in a perfectly natural arrangement, Theridium should be placed near Epeïra, and this genus between the two. There is a true Epeïra, *E. labyrinthea* (my catalogue), which is found in the same locality and which makes a web of the same kind; and I at first suspected that this was a Theridium which had taken possession of the web of that Epeïra, but, besides the character from the legs which does not belong to Theridium, the difference in the cocoons settled my doubts. The cocoons of the Epeïra above mentioned are nearly conical, of an obscure color above, whitish blue beneath; they are hung in a string above the tent. The resemblance of habits in these two species, shows, however, the close affinity between the two genera and this.

### Thomisus, (Walck.).

Eyes 8, generally in two rows bent downward, Pl. 18, fig. 83 or 79; legs variable, but the second generally the longest; lingua contracted at base, wider towards the middle; maxillæ inclined over the lingua. Making no web, but wandering after their prey on flowers, rails, trees, etc. Eight species. This genus, embracing very different species, is not natural. It should include only the Heteropodæ of Walck., which have the two anterior pair of legs sensibly longer than the others. The other species ought to constitute other subdivisions.

## Sphasus, (Walck.). *Oxyopes*, (Latr.).

Eyes 8, unequal in size, Pl. 18, fig. 14 *a*; legs 1. 2. 4. 3. lingua long, rounded at its apex; maxillæ long, narrower at tip. Making no web, except when the female makes her cocoon. Three species. Nothing is known as yet in Europe about the habits of the spiders of this genus, and therefore I will state my observations. There is much [106] similarity between them and the subdivision Sylvaria (Walck.) of Dolomedes, in point of manners and habitus. The three species of Sphasus, known to me, wander in quest of prey about the trunks of small trees or upright sticks, move with great rapidity, and when at rest spread their feet like many species of Thomisus. On the first of September a large female was brought to me in a glass vessel. I call it *Sphasus viridans*. It is of a pale grass color, with the disk of the abdomen yellowish, except an oblong longitudinal line in the middle, which has a double row of three or four oval oblique yellow spots, separated by a longitudinal blackish line; feet pale with yellow joints. Length 0.81 of an inch. It was impregnated and with eggs. After a few days it made a web of very strong threads, like that of Theridium, in the middle of which was placed its cocoon, which is perfectly conical, made with great exactness, and is supplied around with little mammulæ from which depart the threads which bind it to the web. The mother watched it constantly, and never left it as long as she lived. The young were hatched on the 14th of October, and continued together for many weeks during the winter, but gradually died; they were of a deep orange color and full 0.9 of an inch in length. The mother had previously been destroyed by an accident, which I regretted very much, for I have some reasons to think that the young are carried on the back of the mother, as in Lycosa, and wished to have ascertained that fact.

## DOLOMEDES, (Latr.).

Eyes 8, unequal in size, Pl. 18, fig. 73 or 55 ; legs 4. 2. 1. 3. wandering near streams or ponds, often hiding under the surface of the water, or rambling on trees. Six species. Dr. T. W. Harris sent me a species, the female of which constructs a web not unlike that of Tegenaria ; but that retreat is limited to one sex, and probably used only to protect the cocoon until the young are hatched and able to go abroad.

## LYCOSA, (Latr.).

Eyes 8, unequal in size, Pl. 2, Fig. 8 *b*; legs 4. 1. 2. 3. wandering about in quest of prey found under stones, in holes, etc., bearing their cocoons attached to their anus, and carrying their young on their back. Eleven species known to me. Dr. Charles Pickering, of Salem, Mass., presented me with a collection of Araneïdes, in which were six [107] or perhaps seven new species from New England, but which are too much dried up to be well delineated or described. That single fact shows how far this is from being a complete list of North American Spiders. The famous Tarantula of the south of Europe, the bite of which for many years was supposed to produce a disease that music alone could cure, belongs to this genus; and I found on Round Hill, Mass., a species (*Lycosa fatifera*, my catalogue) which is probably very closely related to the European species, and which dwells in holes nearly a foot deep. These holes seem to be dug by the spider, and to be increased gradually, as its size may require : the opening has a ring of filaments woven by the spider to prevent the filling up of the cavity by rain. It is in this genus also that we may witness astonishing instances of maternal tenderness and courage ; and that, too, in the most cruel race of animals, a race in which ferocity renders even the approach of the sexes a perilous act, and condemns every individual to perpetual solitude, and apprehensions of its own kind. When a mother is found with the cocoon containing its progeny, if this be forcibly torn from her, she turns round and grasps it with her mandibulæ. All her limbs, one by one, may then be torn from her body without forcing her to abandon her hold. But if, without mangling the mother, the cocoon be skillfully removed from her, and suddenly thrown out of sight, she instantaneously loses all her activity, seems paralyzed, and coils her tremulous limbs as if mortally wounded; if the bag be returned, her ferocity and strength are restored the moment she has any perception of its presence, and she rushes to her treasure to defend it to the last.

### Attus, (Walck.). *Salticus*, (Latr.).

Eyes 8, unequal in size, Pl. 18, fig. 65; legs usually short and proper for leaping, of different sizes; maxillæ erect, rounded. Wandering in quest of prey, and leaping. Making no web, but tubes of silk for shelter in crevices, under bark, etc.

Twenty-nine species. The numerous species of this genus display skill and varied strategems to seize their prey, which must be interesting to an observer of nature. I have preserved the name of Attus because the name Atta, previously given by Fabricius to a subdivision of Formica, could not be mistaken for this, any more than the Romans would *casus* for *casa*, and a thousand such words. [108]

### Epiblemum, (Mihi).

Eyes 8, somewhat unequal in size, Pl. 18, fig. 59; legs 1. 4. 3. 2. or 1. 4. 2. 3.; lingua short, triangular; maxillæ somewhat pointed above, and a little inclined over the lingua; mandibulæ nearly horizontal, slender, as long as the cephalothorax, tooth as long. Two species. These might be left with Attus, to which they are closely related, but as that genus is large, it needs divisions, and the mandibulæ of these offer a peculiar and striking character, I have concluded to make the 'first of the two following species the type of a new genus. *Epiblemum faustum* obscure, cephalothorax edged with white, with two spots on the disk also white ; abdomen edged at base, and with four short bands, white. *E. palmarum*, deep ferruginous, with two bands on the cephalothorax and the abdomen, white ; second, third and fourth pair of legs whitish.

Besides these, I have three species of Attus, all very small, which have the habitus of Formica ; so much like ants in many respects, that for a long time I neglected to collect them on that account. Their body is elongated, slender, nodose ; and their legs also are slender, either 4. 3. 1. 2. or 4. 1. 2. 3. The cephalothorax in one, and the abdomen in all, are contracted in the middle, so as to give them the appearance of being divided in three or four joints. The other characters coincide generally with Attus. They are found on plants. Should it be thought convenient, those and any other new species with those characters, might be collected under the generic name of Synemosyna.
\* \* \* \* \* \* \* \* \* \* \*

It will be observed, that, in the above arrangement I have departed from that of Latreille in no essential point, but justice requires us to notice, that after the labors of the greatest living entomologist, the method of Walckenaer may still be considered as somewhat more natural than that of Latreille. I have given a sufficient account of the American genera, known to me, to allow any person whose taste may lead him to study this branch, to pursue the subject to a certain extent, and to assist in bringing my Monographia to a less imperfect state than that in which it now is. It is evident to me that if I had correspondents in the various States of this Union who would be willing to send me specimens, I could double my collection in a few years. Some persons have been kind enough to send me several interesting species, particularly Dr. Harris of Milton, and Dr. C. [109] Pickering of Philadelphia, to whom I am much indebted ; but, when stuck through with a pin, and dried as other insects, these become so shriveled as to make it sometimes impossible to recognize them, and always so to describe new species. Spiders should be preserved in diluted alcohol, or brandy, in which they preserve their form, though their colors are usually impaired in it.

The number of one hundred and twenty-five species will appear very large, but I could have swelled the list to one hundred and fifty. Spiders differ from true insects, or at least *winged* insects, in their *growing*. They come out from their eggs very minute, and continue to increase in size, probably for several years in many species; whereas, with few exceptions, insects come out of their *pupa* state, at once, with the size which is peculiar to them. The Araneïdes, in their different ages, present differences of color and marking. The *seasons* also produce a change in the colors of some spiders; and I am nearly convinced that the first frosts produce a total change in the dress of several described Epeïræ which may be referred to one name. These are the considerations which have induced me to be very cautious in adopting new species, and comparing many specimens in different seasons, when possible, before I described them.

.

[From the Am. Journ. Science and Arts, xli, 116.]

Art. XII. DESCRIPTION OF AN AMERICAN SPIDER, CONSTITUTING A NEW SUBGENUS OF THE .TRIBE INÆQUITELÆ OF LATREILLE. By Prof. N. M. Hentz, Florence, Ala.

The genus Aranea of Linnæus, like most of the genera established by that great man, is now in fact an extensive family of the animal kingdom. Walckenaer and Latreille subdivided it, and at once classified the numerous species known to them in an admirable order. We may add the species since discovered, and such subgenera as were not known to those authors, without materially altering their superstructure. But when the work is accomplished, and all nature is described by man, the number of species included in the common word *spider* will be truly amazing. Walckenaer enumerated two hundred and sixty species thirty-four years ago, and Latreille could easily have doubled the catalogue, if the number of species had been mentioned in the last edition of the *Règne Animal*. The writer of this paper, in the course of twenty years, has, at stolen hours, collected and described one hundred and forty-seven species; but he is convinced that *fifty* more could be added; as he has not explored the vast peninsula of Florida, nor any portion of Louisiana. Two hundred species, therefore, would be a low estimate of the number of spiders inhabiting the United States, not including the territories yet unoccupied by civilized men. It is obvious that the number of species throughout the world will amount to more than *two thousand*, when the natural history of all countries is complete. It is equally obvious that the rapidly increasing number of new species requires subdivisions, when it is practicable to make them. The subgenus now proposed is indispensable, as the species cannot be classed under any existing generic name. It will be placed in a natural order immediately after Pholcus.

### Subgenus SPERMOPHORA.

*Eyes*, six, in two clusters, one on each side of the cephalothorax, Pl. 17, fig. 9.

*Legs*, the *first* pair longest, then the *fourth* and *second*, nearly equal, the *third* pair shortest. Length moderate, slender.

*Lip*, wide, triangular.

*Maxillæ*, tapering towards the point, inclined over the lip.

*Mandibulæ*, short, conical, with very small fangs.

The characters derived from the trophi are nearly those of Pholcus, but the absence of the two eyes in front of the cephalothorax, would alone remove this spider from that sub-division. Moreover, the legs, which in Pholcus are excess-ively long, are here of a moderate length. This spider, which is wholly of a pale hue, makes its very loose web in dark places, under rubbish. The female carries in its mandibles its eggs glued together without any silk, until they are hatched. Inhabits Alabama.

This species, the one hundred and thirty-seventh of my MS. catalogue, is there named *Spermophora meridionalis*. Of the one hundred and forty-seven species comprised in this cata-logue, there are not ten mentioned in European works besides those described by Bosc, whose manuscript was never printed.

FLORENCE, ALA., *September* 2, 1839.

[From the Bost. Journ. Nat. Hist., IV, 54.]

Art. VI. DESCRIPTIONS AND FIGURES OF THE ARANEIDES OF THE UNITED STATES. By Nicholas Marcellus Hentz.

(Communicated July, 1841.)

The Publishing Committee think it proper to inform the readers of this Journal, that the following article is the first of a series on the Araneides of the United States, which has been offered for publication, by the author, to the Boston Society of Natural History. These descriptions and figures will be fol-lowed hereafter by others, and the whole will form an illus-trated monograph of all the Spiders observed by Professor Hentz in various parts of this country, and will supply a want [55] which has been long felt in this department of our Natu-ral History.

## Class. ARACHNIDES.
### Order. PULMONARIA.
### Family. ARANEIDES.
### Section. *Tetrapneumones.*
### Genus. MYGALE. Walckenaer.

Characters. *Eyes eight, placed near together on the anterior edge of the cephalothorax, in two rows, variously curved; fang of the chelicera articulated downward; palpi inserted on the extremity of the maxillæ; feet* 4. 1. 2. 3 *or* 4. 1. 3. 2.

*Observation.* The distinction between Mygale and Oletera is artificial, as a slight elongation of the maxillæ of Mygale would place the palpi at the side; witness *Mygale? unicolor.*

### 1. Mygale truncata.

Pl. I, fig. 1. *a.* Eyes. *b.* Trophi. *c.* Side view of the Spider. *d.* Hole in which it resides.

*Description.* Piceous; cephalothorax with a curved impression behind the middle, *chelicera* (mandibulæ) terminated by several points above the fang, hairy; abdomen cylindrical, suddenly truncated at the end, and callous at that place, with concentric grooves and six circular impressions; thighs more or less rufous at base; a white membrane between the joints. Feet 4. 1. 3. 2.

*Observations.* This spider dwells, like other species of this subgenus, in cylindrical cavities in the earth. Though many specimens were found, I never saw the lid described by authors as closing the aperture of its dwelling. The very singular formation of its abdomen, which is as hard as leather behind, and which forms a perfect circle, induces me to believe that it closes with that part, its dwelling, instead of with a lid, when in danger.

*Habitat.* Alabama.

### 2. Mygale solstitialis. [56]

Pl. 1, fig. 2. *a.* Eyes. *b.* Trophi. *c.* Abdomen viewed underneath.

*Description.* Deep black; cephalothorax with two indentations, cheliceres moderately large; abdomen with several impressions above, and four yellow spots underneath; membrane between the joints white; third pair of legs with the third joint short and crooked; feet hairy, 4. 1. 2. 3. A large species.

*Observations.* One specimen only (a male) was found in July, wandering on the ground. The character, derived from the third pair of legs, does not seem to be a mere sexual distinction, as *Mygale carolinensis*, the next species, has the same peculiarity, and the description was taken from a female.

*Habitat.* Alabama.

### 3. Mygale carolinensis.

Pl. 1, fig. 3. *a.* Eyes.

*Description.* Brownish, very glossy; cephalothorax with two slight impressions near the base; abdomen blackish, not glossy; third joint of the third pair of legs very short and crooked; feet 4. 1. 3. 2.

*Observations.* This species was communicated to the author by the late Mr. Levi Andrews, of Chapel Hill, North Carolina, a promising young naturalist, snatched by consumption from his numerous friends, and to the memory of whom this tribute is due. The character derived from the third pair of legs is not a sexual one, as this was a female, and the description of *Mygale solstitialis* was taken from a male, which has the same character.

*Habitat.* North Carolina.

### 4. Mygale gracilis.

Pl. 1, fig. 4. *a.* Eyes. *b.* Right palpus, with the maxilla.

*Description.* Rufous; cephalothorax somewhat six-sided, long and narrow; abdomen plumbeous, two nipples very long;

feet long, hairy, penultimate joint of the anterior pair with a notch ; feet 4. 1. 2. 3. [57]

*Observations.* This spider, hitherto always found in mid-winter, under stones or on the ground, is probably not the male of *Mygale carolinensis* ; but the peculiarity of its *first* pair of legs, is, no doubt, a sexual character. The same joint of the feet of the male of my *Dysdera bicolor*, is not only bent, but has powerful prongs and bristles, which nature has given him as a defence, or as the means of grasping the female.

*Habitat.* Alabama.

### 5. Mygale? unicolor.

Pl. 1, fig. 5.  *a* and *b.*

*Description.* Deep rufous ; cephalothorax depressed in the middle, with two impressions, cheliceres very large ; abdomen smooth ; third pair of legs with short, very thick joints ; feet 4. 1. 2. 3.

*Observations.* This species is very distinct from any other, particularly by the manner in which its palpi are inserted. Were the maxillæ extended a little more at their extremity, this spider should be placed in the sub-genus Oletera, which follows. The specimen from which this description was taken (a female), was turned up by the plough in a field, in the month of May. The manner in which the spiders belonging to Mygale and Oletera live, hidden under ground, and probably issuing out only at night, prevents our becoming acquainted with their habits. I doubt whether the males ever dwell in tubular habitations. Much remains yet to be done to complete the history of this genus and that of the next.

*Habitat.* Alabama.

[From the Bost. Journal Nat. Hist., IV, 223. Art. XVI.]

Genus. ATYPUS, Latr. *Oletera*, Walck.

Characters. *Cheliceres large with a fang nearly equal to their length, articulated downward; maxillæ tapering upward, insertion of the palpi lateral ; lip concealed ; eyes eight, subequal, collected in front of the cephalothorax, two in the centre, and on each side of these there is a cluster ; feet* 4. 1. 2. 3.

*Habits.* Araneides sedentary, dwelling in silk tubes placed in the ground. [224] *

*Observations.* The habits of the animals of this subgenus are but little known, owing to the obscure locations which they select. They are probably nocturnal.

**Atypus niger.**

Pl. 2, fig. 1. *a* and *b*.

*Description.* Deep black; cephalothorax flattened, horny, with three depressions ; a white membrane at the base of the cheliceres. A small species.

*Observations.* A solitary individual (a male) was found in June, on newly turned soil, at Northampton, Mass., by the son of the late Prof. W. D. Peck. I am not acquainted with *A. rufipes* found by Mr. Milbert, near Philadelphia.

*Habitat.* Massachusetts.

[Marietta. O., ♂ Wm. Holden. J. H. E.]

Genus. DYSDERA. Latr. Walck.

Characters. *Cheliceres large, fangs articulated inward ; maxillæ straight, wide at base, narrowed above the insertion of the palpi, inner edge cut obliquely towards the point ; lip half as long as the maxillæ, emarginate at tip ; eyes six, subequal, four in a line curved towards the base, and one each side nearer the anterior edge, but leaving an open space between them ; feet, first pair longest, the second and fourth nearly equal, the third shortest.*

*Habits.* Araneides sedentary, dwelling in silken tubes, under stones or in crevices.

*Observations.* The large size of the cheliceres, and other minor characters, show some affinity to Mygale. The only species here described being made known to me by Dr. T. W. Harris, of Massachusetts, I am not acquainted with many facts necessary to give a good history of this subgenus.

### Dysdera interrita.

Pl. 2, fig. 1.  *a* and *b*.

*Description.* Ferruginous ; cephalothorax and trophi piceous.

*Observations.* This species was communicated to me by my excellent friend Dr. T. W. Harris, of Massachusetts, who [225] sent me the male and the female, also, with a correct sketch of both sexes. It inhabits that State, and was found in cavities under ground, under rotten wood, etc., in the month of May.

[♀, length 12.4 mm.; cephalothorax 5.2 mm.; legs 13. 12, 9.5, 12.8.

♂, " 14.5 mm. ;  " 4.6 mm.; " 12.5, 11.8, 9, 11.8.

Palpus of ♂, Pl. 20. fig. 1. Claw of first foot. fig. 1a. Young with cephalothorax reddish-yellow like the legs. This is the only species of Dysdera I have found in Massachusetts. Hentz's original drawings are lost.

*Dysdera rubicunda* Blackwall, Spiders of Gt. Britain and Ireland.

"  "  Menge. Preussische Spinnen.

West Roxbury. Mass.  ♂ and young in June, ♀ in October. F. G. Sanborn.

Malden. Mass.  H. L. Moody.

Brookline, Mass.  A. Smith.

Mass.  ♀, Wm. Holden.  J. H. E.]

### Genus. PYLARUS. Mihi.

*Characters.* *Cheliceres small, fang very short, maxillæ slightly inclined over the lip, long and slightly rounded at tip; lip tapering, half as long as the maxillæ; eyes six, equal, in three pairs, two in the middle and two each side, placed diagonally on a common elevation, nearer the anterior edge; feet, first, second and fourth pairs subequal, third shortest, penultimate joint of the first pair armed with hooks in the male.*

*Habits.* Araneides sedentary, forming a silken tube in crevices of walls, with a few threads spreading from the orifice unto the edge of the crevice, the spider watching near the entrance with its three anterior [pairs of] legs extended out.

*Observations.* This subgenus, which was first confounded by me with Dysdera, differs from it by the small size of its cheliceres, and the position of its eyes. By the habits of the spiders which compose it it bears close affinity to Segestria, but the position of its eyes is reversed. It is obvious that as this is not Segestria, and cannot be referred to Dysdera, it must constitute a new subgenus.

### 1. Pylaris bicolor.

Pl. 2, fig. 3. ♀. *a* and *b*. Fig. 4. ♂. *a*. palpus.

*Description.* Cephalothorax piceous; abdomen bluish-black : first and second pairs of legs blackish, hairy, third and fourth piceous. Male piceous; abdomen with the base and sides paler; feet with few hairs, penult joint of the first pair crooked and with two strong spines, the antepenult with about four strong bristles on each side.

*Observations.* This spider, which is very common in Alabama, makes its tubular habitation in the crevices of walls, commonly waiting near the orifice with its three first pairs of legs directed forwards. Its silken tube spreads out on the outside, [226] and whenever an insect touches one of the threads the spider issues out with the rapidity of a hawk and seizes its victim, which it carries immediately within. In damp, rainy nights, the males and females are often found wandering from their homes. The male, which is provided with very unusual means of defence on its first pair of legs, is nevertheless excessively cautious in his approach to the residence of the female. He advances with the utmost caution, remaining motionless near the entrance for hours. This takes place in October. I once observed a male in that situation, and wishing to secure him, suddenly transfixed his cephalothorax with a pin, when the female furiously rushed out and boldly grasped him, struggling to carry him off; and she nearly succeeded in robbing me of my prey, which she seemed to consider her own. I have found this species hibernating in silken tubes, along with various species of Attus, in December and January. This proves that *Dysdera pumila* is not a variety of it.

*Habitat.* North Alabama.

[♀, length, 9 mm.; cephalothorax, 4 mm.; legs, 8.6, 8.5, 6.2, 8. Front foot with hairs removed, Pl. 20, fig. 2. Cephalothorax and legs yellow brown, darkest toward the head. Abdomen purplish brown, lighter at the sides. Young lighter, with cephalothorax and legs yellow. The original drawings lost. Salem, Mass., July 10, ♀, in old mud cell of wasp under a stone, with cocoon of thirty-four eggs. Another ♀ in closed silk tube, with cocoon of eggs. Providence, R. I., Oct. 29. Young and old in thick silk tubes under bark. Mayport, Fla.; Ohio. Wm. Holden. J. H. E,]

### 2. Pylarus pumilus.

Pl. 2, fig. 5.

*Description.* Livid yellow; abdomen dusky on the disk and towards the apex; first and second pairs of legs with the two last joints dusky; hairy.

*Observations.* This species is usually found under the bark of trees, enclosed in silk tubes.

*Habitat.* North Carolina, North Alabama.

[Probably young of *P. bicolor.* J. H. E.]

### Genus. FILISTATA. Latr.

*Characters. Cheliceres small, incapable of reciprocal motion, fang very small; maxillæ bent and surrounding the lip, terminating in a point; lip more than half the length of the maxillæ, widest in the middle, ending in a point; eyes eight, subequal, placed closely together on a common elevation, [227] two in the*

*centre, usually black, three on each side, leaving a space above
and below opened towards the middle ones; feet,* 1. 4. 2. 3.

*Habits.* Araneides sedentary, forming a tube of silk in the
crevices of old walls, with loose threads spread out round the
orifice, the spider usually watching at the entrance.

*Observations.* The characters derived from the cheliceres,
which are articulated together so as to allow of little or no
reciprocal motion, is peculiar to this subgenus. On the whole,
it seems to have a greater affinity to Clotho than to any of the
Tetrapneumones of Latreille; and, by its habits, it is closely
related to my Pylarus and to Segestria. Independent of the
difficulty of ascertaining the pulmonary orifices, these points of
affinity between Dipneumones and Tetrapneumones show that
the distinction may prove an artificial one.

### 1. Filistata hibernalis.

Pl. 2, fig. 6. ♀. *a.* Trophi, with the palpi of the male. · *b.* Eyes.

*Description.* Deep mouse-colored, covered with fine short
hair; cephalothorax darker; cheliceres small. Male, pale gray
or livid; palpi excessively long, two middle eyes black, the
others shining white.

*Observations.* It makes a tubular habitation of silk in crevices on old walls or rocks, throwing an irregular web which is
spread on the wall or stone around the aperture. It comes out
occasionally during the winter, but cold is apt to render it torpid, and it then remains several days in the same situation,
moving slightly in the middle of the day. In walking, it uses
its palpi like feet, and these organs are very long, particularly
in the male. I saw one of this species change its skin in con-
finement. It had previously lost a leg by some accident, but
after moulting it had a new one which had all its joints, only a
little shorter than the natural size; its cocoon is spherical.

*Habitat.* South Carolina on the sea-coast, North Alabama
on the banks of the Tennessee.

[♂, length, 13.8 mm.; cephalothorax, 6.2 mm.; legs, 23, 19, 15.6, 20.6.
Fernandina, Fla., Sept. 1, with cocoon of young. E. Palmer. J. H. E.]

## 2. Filistata capitata.

Pl. 2, fig. 7.  ♂.

*Description.* Dusky brown; eyes much elevated, cephalo-thorax with a deep longitudinal impression, beginning above the eyes and not reaching the base; cheliceres not closely articulated together: abdomen and feet with short hairs.

*Observations.* This species, communicated to me by Mr. Thomas R. Dutton, was brought by him from Georgia, where it inhabits crevices like *Filistata hibernalis.* No females were brought. It is strange that its cheliceres are not joined together as in that species. The trophi in other respects correspond entirely with it.

*Habitat.* Georgia.

[Mayport, Florida.  ♂.  Wm. Holden.  J. H. E.]

## Genus.  Lycosa.  Latr.

*Characters.* *Cheliceres large, fangs moderate; maxillæ short, parallel, cut obliquely at the tip : lip short, slightly emarginate at the upper edge, which is slightly narrower than the base; eyes eight, unequal, four small placed anteriorly in a straight or slightly curved line, two large placed above the two external ones of the first line, two of middle size placed further out towards the base and nearly forming a square with the intermediate ones; feet, 4. 1. 2. 3.*

*Habits.* Araneides making no web, wandering for prey, hiding under stones and frequently making holes in the ground in which they dwell, making at the orifice a ring of silk, forming a consolidated entrance; cocoon usually orbicular, often carried about by the mother, the young borne on the back of her abdomen.

*Observations.* The subgenus Lycosa is not variable in its characters like Dolomedes. The lower row of eyes is straight in some species and more or less curved in others, but I could not avail myself of this to make any satisfactory subdivision.

The upper mammulæ, it is true, are longer in *Lycosa lenta*,
but I found them to vary in length in others so imperceptibly
that I could not adopt any of the three families [229] of
Walckenaer, which appear to me quite artificial. These spiders are the eagles and lions of the family. They are found
swarming on the ground, running with great agility, a property
belonging to those spiders in which the fourth pair of legs is
longest. Most are usually found wandering for prey, except
when engaged in maternal duties : others dwell in holes several inches deep, well rounded and supplied with a ring of silk
and little straws, consolidated so as to prevent the crumbling of
the earth. I have found one of these in the winter which was
supplied with a lid, and probably they all close the orifice for
hibernation. The mother carries its cocoon attached to the
posterior part of the abdomen. Small species ramble about
with these ; but the larger ones watch them in their habitation
or under stones. The moment the young ones are hatched
they climb on the abdomen of the mother and remain there for
a considerable time. They give a monstrous and horrible appearance to the mother, which seems hairy, and twice as large
as usual. If the parent be touched, or forcibly arrested, the
young spiders instantly disperse and disappear. The mother
when deprived of its cocoon, seems to lose all her ferocity and
activity, but if it be placed near her, the moment she perceives
it these powers return, and she rushes to the cocoon, which she
grasps with renewed vigor. She defends her progeny to the
last, and her feet can be torn from her one by one, before she
can be compelled to abandon her treasure. Thus can maternal
tenderness be exhibited in beings which are relentless to their
own species, and even to the sex which gives life to its progeny.
It is extremely difficult to distinguish the different species of
Lycosa, owing to the infinite varieties in colors, marking and
size. Future writers will probably clear the confusion which I
boast not of having removed during twenty years of studious
attention to this subgenus.

## 1. Lycosa fatifera.

Pl. 2, fig. 8. *a, b.*

*Description.* Bluish black; cephalothorax deeper in color at the sides; cheliceres covered with rufous hairs and with a red elevation on their external side near the base; one of the largest species. **[230]**

*Observations.* This formidable species dwells in holes ten or twelve inches in depth, in light soil, which it digs itself; for the cavity is always proportionate to the size of the spider. The orifice of the hole has a ring, made chiefly of silk, which prevents the soil from falling in when it rains. This Lycosa, probably as large as the *Tarantula* of the South of Europe, is common in Massachusetts; but we have not heard of serious accidents produced by its bite. Its poison, however, must be of the same nature and as virulent. The reason perhaps why nothing is said of its venom, is, that so very few instances can have occurred of its biting any body. All persons shun spiders, and these shun mankind still more. Moreover, their cheliceres cannot open at an angle which can enable them to grasp a large object. Without denying its power to poison, which it certainly has, it is well to expose popular errors, such as that of the Romans in regard to the bite of the shrew, which it is now proved cannot open its mouth wide enough to bite at all. This spider, when captured, shows some combativeness, and has uncommon tenacity of life. It is a laborious task to dig down its deep hole with the care necessary not to injure it. I have at times introduced a long slender straw downward, till I could feel a resistance, and also the struggle of the tenant; and I could perceive that it bit the straw. In one or two instances, by lifting the straw gradually, I brought up the enraged spider still biting the inert instrument of its wrath. It probably lives many years. A piceous variety is found in Alabama, with the two first joints of the legs, pectus and abdomen yellowish underneath, or lighter in color.

*Habitat.* Massachusetts, North Alabama.

### 2. Lycosa (Tarantula) Carolinensis ? Bosc. M. S.

Pl. 2, fig. 9.

*Description.* Mouse-colored; cephalothorax with an in-dented blackish mark at base; cheliceres covered with rufous hairs in front, and with a red elevation; abdomen with several whitish dots and angular transverse lines on the disk, sides nearly white; beneath, usually quite black, except the legs, [231] which are whitish, the joints tipped with black. Male with nearly the same marks, very black beneath. Attains a very large size.

*Observations.* This spider has the same habits as *L. fatifera*, making deep excavations in the ground. It is frequently found under stones, and possibly it is in such places, nearer the sur-face, that the eggs are hatched. The female carries her young on her back, presenting a hideous aspect, being then apparently covered with animated warts. The little monsters have the instinct, if the mother is much disturbed, to escape and scatter in all directions. The male, not unfrequently of an enormous size, is often found wandering in October and November, in Alabama, and sometimes enters houses.

*Habitat.* North Carolina, Georgia, North Alabama.

[♂, cephalothorax, 13.2 mm.; legs, 31, 28, 27, 36.
♀,     "     10.4 mm.;  "   33, 30, 26.6, 35.2.
Palpus of ♂, Pl. 18, fig. 3.
Essex County, Mass., ♂ and ♀.
Worcester, Mass., October, ♂. F. G. Sanborn.
Mt. Desert, Me. S. Henshaw.
Ohio, ♂, ♀. W. Holden. J. H. E.]

---

[From the Bost. Journ. Nat. Hist., IV, 386. Art. XXXI.]

### 3. Lycosa lenta.

Pl. 3, figs. 1-4.

*Description.* Piceous, hairy; cephalothorax with a waved fascia of a dark color, and several pale marks. Abdomen with

two longitudinal rows of indistinct black spots above, beneath
with a large black spot, with a yellowish mark in the centre.
A pale variety occurred in North Carolina, without the yellow
mark. [387]

*Observations.* This common and powerful species is found
wandering in fields, attacking and subduing very large insects.
The female carries her young on her back, which gives her a
horrible appearance. If caught or wounded, the little ones
escape rapidly in all directions; but the mother is faithful to
her duties, and defends her progeny while life endures. It
hides under stones, logs, etc.

*Habitat.* Pennsylvania, North and South Carolina, etc.

### 4. Lycosa ruricola.

Pl. 3, figs. 5, 6.

*Description.* Pale or livid testaceous; cephalothorax with
black marks, two large ones at base; cheliceres black with
yellow hair at base; abdomen varied with black marks and pale
dots above, a large black spot underneath; feet with indistinct
livid rings.

*Observations.* A male and a female of this species were
found with a white spot in the middle of the black one on the
venter, but as the marking differed somewhat from the above,
they may constitute a different species. They are always
found wandering on the ground.

*Habitat.* Carolina, Alabama. October, November.

### 5. Lycosa saltatrix.

Pl. 3, fig. 7.

*Description.* Piceous; cephalothorax with two darker longitudinal bands; abdomen plumbeous or mouse-colored, with four
dark points and a pale longitudinal line; legs hairy, with many
dark bands. Male inclining to a rufous tinge.

*Observations.* This small spider, first found in South Carolina, runs about on the ground, the female carrying her cocoon

attached to the hinder part of her abdomen. When deprived
of it she remains near; and, if allowed, she grasps it in her
cheliceres and carries it off. The cocoon, of a slate color, is
orbicular, and contains about fifteen eggs. This is probably
[**388**] related to *Aranea saccata* of Europe. I have found in
Alabama a spider, which may not differ specifically from this,
which was larger and of a mouse color, with very indistinct
markings, except its legs, which agreed with the drawing ac-
companying this. Its cocoon. which it carried in the usual
way, was also of a bluish pale slate color, but it was lenticular,
being composed of two concave plates of strong texture, united
loosely at the edge; and it contained about sixty yellow eggs —
notwithstanding the apparent difference, I refer it to this species. It is probable, however, that future naturalists will de-
fine two or more species, which I may have confounded or
referred to this description.

*Habitat.* The United States.

### 6. Lycosa erratica.

Pl. 3, fig. 8.

*Description.* Brown or piceous; cephalothorax with one
longitudinal blackish line each side; abdomen with a forked
longitudinal fascia and several spots black, a large black spot
underneath, sometimes a white spot surrounded with black;
male the same.

*Observations.* This species, which becomes very large, I
formerly supposed to be a variety of *L. lenta*; but it was always found wandering and never in holes; I therefore consider
it as perfectly distinct, having been often seen, generally run-
ing in the grass.

*Habitat.* Massachusetts, Alabama.

[Marietta, O. ᵢ. Wm. Holden. J. H. E.]

**7. Lycosa litoralis.**

Pl. 3, fig. 9.

*Description.* Livid white; cephalothorax varied with livid gray markings; abdomen with a pale waved fascia; feet and palpi with some hairs, and with pale gray rings on all joints, 4. 1. 3.¯2., the 1st visibly longer than the 3d, the 3d full as long if not longer than the 2d. [389]

*Observations.* This distinct species is always found near water under boards, leaves, stones, etc., moving chiefly by jumps when escaping. Often observed in the same localities.

*Habitat.* North Carolina. April.

[Marietta, O. ♀. Mayport, Fla. ♂. Wm. Holden. J. H. E.]

**8. Lycosa maritima.**

Pl. 3, fig. 10.

*Description.* Pale yellow, almost white; cephalothorax with faint indented lines; abdomen with two longitudinal rows of dots of a pale hue.

*Observations.* This spider was found on the beach of Bear Island in the bay of St. Helena, South Carolina. Dr. Charles Pickering sent me one from Salem, informing me that it is common in Massachusetts. It runs with great speed on the sand still wet with the ebbing water of the ocean.

*Habitat.* South Carolina, Massachusetts, and probably all the Atlantic coast.

**9. Lycosa aspersa.**

Pl. 3, figs. 11, 12.

*Description.* Greenish obscure; cephalothorax dark, obscure, varied with black marks and a few red lines about the eyes; chelicerae very large; abdomen obscure, with small black spots in three rows, varied with yellow and black in wrinkles underneath, feet with black rings.

*Observations.* Though it is excessively difficult to distinguish between species and varieties in this subgenus, yet I must consider this as distinct from *L. riparia.* It was found on a barren hill at a great distance from water.

*Habitat.* Alabama. September.

## 10. Lycosa riparia.

Pl. 3, figs. 13–15.

*Description.* Brownish or greenish black; cephalothorax varied with blackish, with a narrowed yellowish line which [390] reaches the trophi; abdomen above with triangular black spots more or less interrupted, and a row on each side of whitish dots more distinct towards the apex, a tuft of black and of white hairs at base in both sexes; beneath testaceous or yellow, speckled with dots, and a line and two spots near the base sometimes wanting, black; feet with black or greenish brown rings. 4. $\overline{1.\ 2.\ 3.}$ In the male the two rows of white dots on the abdomen are arranged in the form of interrupted lines, and the rings are obsolete on the feet, which are long, slender and hairy.

*Observations.* This common spider is aquatic in its habits, always found near or on water, and diving with ease under the surface, when threatened or pursued.

*Habitat.* North Carolina, Alabama. All seasons.

## 11. Lycosa punctulata.

Pl. 3, figs. 16, 17.

*Description.* Pale rufous; cephalothorax whitish or yellowish, with four longitudinal blackish lines; abdomen whitish or yellowish, with a longitudinal band, blackish, whitish underneath, with many black dots.

*Observations.* This spider, captured at Germantown, was communicated to me by Dr. Charles Pickering. It was found also in Alabama, in November, agreeing in every respect with the description. It was a male also. A female was found

September 28th agreeing with the design. The species is therefore well established.

*Habitat.* Pennsylvania.

[Rushville, O. Wm. Holden. J. H. E.]

**12. Lycosa scutulata.**

Pl. 4, figs. 1, 2.

*Description.* Testaceous; cephalothorax with one longitudinal band and one line on each side, blackish; abdomen with a longitudinal broad band, blackish, with about four diagonal spots, and a narrow edge, each side of it yellowish; [391] same color underneath, with very minute black dots on the abdomen; legs brownish with some blackish lines. Male with the first pair of legs mostly black, and part of the fourth pair also black underneath.

*Observations.* This common and very distinct species attains a very large stature. It is most commonly found wandering in quest of prey, and like *Lycosa saltatrix*, carries its cocoon attached to the abdomen behind. The cocoon is very large, spherical and whitish, containing from one hundred and fifty to two hundred eggs, which hatch before the cocoon is opened. The yellow spots on the abdomen seem to be wanting in the young.

*Habitat.* Alabama.

[Ohio. ♂, ♀. Wm. Holden. J. H. E.]

**13. Lycosa sagittata.**

Pl. 4, figs. 3, 4.

*Description.* Yellowish brown; cephalothorax with a pale longitudinal band : abdomen dusky also, with a pale band with angular edges, whitish underneath, with minute black dots and two curved black bands which join together at base and at the apex where they spread out: pulmonary region brownish: feet varied with blackish.

*Observations.* This species is distinct from any other. It was found wandering, and seems to be rare.

*Habitat.* North Alabama, April.

### 14. Lycosa ocreata.

Pl. 4, fig. 5.

*Description.* Obscure; cephalothorax, with a broad, pale longitudinal band, with a blackish edge; abdomen blackish at base, the black spreading each side, with a few black dots each side towards the apex; feet varied with brown or blackish; antepenult joint of the first pair large, black and hairy, the intermediate one and the thigh black at tip; feet, 4. $\overline{1}$. 2. 3. A male.

*Observations.* This species is not rare, in meadows, near water.

*Habitat.* North Carolina. [392.]

[Marietta, O. ♂. Wm. Holden. J. H. E.]

### 15. Lycosa venustula.

Pl. 4, figs. 6, 7.

*Description.* Cephalothorax yellowish, with two bands and edge black; abdomen piceous, paler in the middle towards the base, with a row of abbreviated black lines approximating towards the apex, pale gray underneath, with a row of minute black dots each side approximating towards the apex; feet rufous. A middle size species.

*Observations.* This spider is common on the ground, but inasmuch as only males are found, it is likely it will ultimately be referred to some other species; which, I cannot tell.

*Habitat.* Alabama. April.

### 16. Lycosa milvina.

Pl. 4, fig. 8.

*Description.* Pale yellowish; cephalothorax varied with brownish; abdomen brownish with a scalloped band, widening towards the base, and two lateral spots yellowish, pale yellowish spotless underneath; feet varied with brownish, hairy, particularly the third and fourth pair. A small species.

*Observations.* This is a very distinctly marked species, which occurred only once.

*Habitat.* Alabama. September.

**17. Lycosa saxatilis.**

Pl. 4, figs. 9, 10.

*Description.* Pale bluish; cephalothorax varied with grayish; abdomen grayish or blackish, with pale bluish spots or dots, pale gray underneath; feet long and slender, hairy, with many black rings. 4. 2. 1. 3. or 4. 2. 3. 1. A small species.

*Observations.* This slender little *Lycosa* is a very distinct species. It runs with surprising agility and swiftness. It was found in the mountains of North Alabama.

*Habitat.* Alabama. August. [393.]

**18. Lycosa funerea.**

Pl. 4, fig. 11.

*Description.* Cephalothorax blackish; abdomen with four approximate spots and four bent lines yellowish; feet varied with rufous and blackish. A small species.

*Observations.* This species abounds on the ground. It has the habitus of a *Herpyllus*, and runs with great rapidity. The male and the female were often found agreeing with the description.

*Habitat.* Alabama. May.

Genus. CTENUS. Walck.

Characters. *Cheliceres large, fangs moderately large; maxillæ short, parallel, cut obliquely at tip: lip about half the length of the maxillæ, pointed; eyes eight, unequal, in three rows, two eyes of middle size form the lowest row, intermediate row composed of four eyes, the two middle ones largest, the two external ones smallest ; last row formed of two large eyes, borne on tubercles and placed farther apart than those of the middle row ; feet, fourth pair longest, then the first, then the second, the third being shortest.*

*Habits.* Araneides wandering for prey, making no web for a dwelling.

*Observations.* This subgenus seems to be related to Lycosa and Dolomedes.

### 1. Ctenus hybernalis.

Pl. 5, figs. 1–4.

*Description.* Deep rufous ; cephalothorax black above with a longitudinal yellowish band ; abdomen black, with a serrated longitudinal yellow band above, and with four diagonal lines of minute yellow dots beneath.

*Observations.* This was found in a cavity in the ground in the month of January.

*Habitat.* South Alabama. [394.]

### 2. Ctenus punctulatus.

Pl. 5, figs. 5, 6.

*Description.* Yellowish rufous ; cephalothorax with two longitudinal blackish lines and two fainter scalloped ones on each side ; abdomen with two subobsolete lines of minute white dots, becoming more distinct towards the apex, where may be seen a few irregularly placed white dots on the outside of the lines, same color unspotted beneath ; feet, 4. 1. 2. 3. or 4. 1. 3. 2.

*Observations.* This spider was found at the foot of a tree in a moist place near a mountain stream, running through a .forest.

*Habitat.* Alabama. August, September.

### Genus.  DOLOMEDES.  Latr.

Characters.  *Cheliceres moderately large; maxillæ short,
parallel, somewhat wider above the insertion of the palpi; lip
short, suborbicular; eyes eight, unequal, in two rows, the anterior
one slightly curved, the posterior one wider and much curved from
the base towards the anterior one; exterior eyes borne on tubercles;
feet, the fourth, second, and first pair are nearly equal, the third
being the shortest.*

*Habits.*  Araneides wandering after prey, making no web,
except during the rearing of the progeny, hiding under stones,
sometimes diving under water; cocoon usually orbicular, carried by the mother.

*Observations.*  The subgenus Dolomedes is the link between
Ctenus and Lycosa, and its characters are somewhat variable.
In the first tribe (the Arboreæ), which differ wholly from the
Sylvaria of Walckenaer, the arrangement of the eyes is almost
that of Lycosa ; and in the Ripuariæ the arrangement of the
eyes leads to Micrommata.  The spiders of this genus differ in
their habits also ; those of the two first tribes dwell on trees,
or in cavities; those of the third are found near water, and
run on its surface with great rapidity ; they can even dive, and
have recourse to this when in danger. [395] Several, perhaps
all species, construct on bushes a web somewhat like that of
Agelena, for the protection of the cocoon, and the rearing of
the young.  This is another resemblance to Micrommata.

*Order of the species* DOLOMEDES.

Tribe I.  ARBOREÆ, *middle eyes much larger than the rest.*
Tribe II.  TENEBROSÆ, *eyes subequal, lower row as much
curved as the upper.*
Tribe III.  RIPUARIÆ, *eyes subequal, lower row straight or
slightly curved.*

(*Arboreæ.*)

### 1. Dolomedes tenax.

Pl. 5, fig. 7.

*Description.* Grayish; cephalothorax edged with black, varied with blackish on the disk; abdomen also edged with black near the base, varied longitudinally, with blackish on the disk, about three whitish dots on each side near the apex, pale beneath, with two obscure longitudinal lines; feet, with blackish bands above, pale beneath. 4. 2. 1. 3. Never large.

*Observations.* This distinct species is always found on upright sticks, small trees, etc., turning round to avoid an attack in the same manner as *Oxyopes scalaris*, which it resembles so much that for a time I could not distinguish one from the other. It spreads its feet like Thomisus. The form of its cephalothorax is peculiar, the *head* being elevated and well-defined from the thorax. It must not be taken for the young of *D. tenebrosus*, which resembles the old, and dwells in dark places, whereas this is quite a diurnal species, fond of broad daylight.

*Habitat.* North Carolina.

### 2. Dolomedes hastulatus.

Pl. 4, fig. 9.

*Description.* Pale or greenish gray; cephalothorax varied with blackish; abdomen with a blackish band, with rounded [396] edges near the base, and terminating with a hastate point towards the apex; feet varied with gray or blackish. 2. 4. 1. 3.

*Observations.* This was found in September in a web, like that of Agelena. This can be readily distinguished from *D. tenax*, particularly by the form of its cephalothorax, in which the head is not elevated as in that species. The second pair of legs being sensibly the longest. This species could almost be referred to Micrommata. It was found in March upon the stump of a tree not far from a stream.

*Habitat.* Alabama. September.

(*Tenebrosæ.*)

### 3. Dolomedes tenebrosus.

Pl. 5, figs. 10, 13.

*Description.* Livid brown; abdomen and cephalothorax varied with blackish angular markings; feet annulated with blackish; frequently measuring over four inches from the extremity of the first pair of legs to that of the fourth pair; male with legs 1. 2. 4. 3.

*Observations.* This spider, one of the largest of the whole family, is very common in dark, retired places, hiding in crevices during the day, and issuing at night from its retreat for the purpose of seeking for prey. It does not seek the vicinity of water near which it was never seen, but dwells generally in elevated dry places. The female does not make a web, but carries its cocoon, grasped with her cheliceres. The cocoon is orbicular whitish, and of the size of a common cherry. I have occasionally seen this Dolomedes in the daytime, but it seemed always inactive, and easily captured. It can be readily distinguished from *D. albineus*, by its having no yellowish spot under the abdomen, and by the white hairs on its legs.

*Habitat.* Carolina, Alabama, Massachusetts?

[Ohio. ♀, ♂. Wm. Holden. J. H. E.]

### 4. Dolomedes scriptus.

Pl. 6, fig. 1.

*Description.* Pale brownish; cephalothorax varied with black and white : abdomen with a broad blackish band intersected by waved white lines, and usually edged with whitish, pale spotless underneath; feet varied with obscure brown, ultimate joint tipped with blackish.

*Observations.* This species was found in great numbers on the margin of a stream under stones. The two triangular black spots, visible on the cephalothorax of *D. urinator* and *D. lanceolatus*, are obsolete on this. Many were examined,

and agreed with this, only the white edge of the band being
less distinct in some.

*Habitat.* Alabama. March.

### 5. Dolomedes albineus.

Pl. 6, fig. 2.

*Description.* Mouse-colored ; abdomen varied with angular
markings above, beneath with a yellowish longitudinal band,
edged with black ; feet with alternate black and white rings,
the white rings formed by long white hairs ; the legs have also
a few black bristles : male with legs 1. 2. 4. 3. As large as
*D. tenebrosus* nearly.

*Observations.* This species which, at first sight, might be
taken for *D. tenebrosus*, does not dwell habitually in caves and
cellars, but is usually found on the trunks of trees, yet in dark,
shady places. Several females were found, and a male, also.
One of those females was captured by a child, who transfixed her
cephalothorax with a pin. Finding she was full [190] of eggs,
I was desirous to see whether she could survive the wound. I
placed her in a glass jar, and, according to my expectations,
nature made an effort, that she might live for the protection of
her progeny. The wound, which in other cases, would have
proved immediately mortal, healed readily, and after remaining
inactive about three days, she made a cocoon of a light brown
color, and orbicular, in which her eggs were placed. She held
it constantly grasped in her cheliceres, and seemed intent on
watching it to the last, but the effort being made, her strength
failed ; the wound opened again, and the fluids running out, she
very gradually lost all her muscular power, but faithful to her
duties, the last thing which she held was the ball containing
her future family. Can maternal tenderness be more strikingly
exhibited ?

*Habitat.* Alabama.

[Pl. 18, fig. 73, eyes. The area of the eyes is black ; thighs
and breast all shining piceous underneath. Taken July 3.
*Supplement.*]

*(Ripuariæ.)*

### 6. Dolomedes urinator.

Pl. 6, fig. 3.

*Description.* Livid brown, somewhat hairy; cephalothorax with obscure marks uniting towards the centre, two approximate wedge-like black spots on the disc; a black spot behind the external posterior eyes; abdomen varied with curved blackish lines, and with eight, ten, or twelve white dots surrounded with black; feet with brownish rings.

*Observations.* This large spider is found near water, on which it runs with great swiftness. When closely pursued, it dives under the surface, and conceals itself under some leaves or rubbish till danger is past. It is to this species, probably, though possibly to *D. lanceolatus*, that must be referred a sketch sent me by Dr. T. W. Harris, with the description of the web made by the female. I have not yet been fortunate enough to find the web of either species. I have found a specimen of this, on the 3d of March, in Alabama, basking in the sunshine on the south side of a tree, on the margin of the Cypress creek. I took it for *Micrommata carolinensis*, notwithstanding its large size, but its markings corresponded [191] entirely with this; only its general color was bordering on testaceous or pale brown.

*Habitat.* North Carolina, Alabama.

[New Lexington, O., ♀. Wm. Holden. The sketch here mentioned is published in Harris' Correspondence, p. 4, and a description of the nest on p. 37. It was found in Milton, Mass., August, 1824. J. H. E.]

### 7. Dolomedes lanceolatus.

Pl. 7, fig. 12.

*Description.* Brownish; cephalothorax with two approximated triangular black spots, and a yellowish band round the disc which does not extend to the margin, and is interrupted at base; abdomen with a yellowish band, which has on each side two branches directed towards the disc; feet varied with pale rings.

*Observations.* This spider is always found near, or on water, running on it with surprising agility, preying often on large

aquatic insects. A female of Dolomedes was twice found on
high bushes by my friend, T. W. Harris, in Milton, Massachusetts, "on a large, irregular, loose, horizontal web, at one extremity of which was situated her follicle, or egg-bag, covered
with young. The parent appeared watching them at some
distance." This spider can dive and stay a considerable time
under water, to avoid its enemies. It was found in March, in
Alabama, under stones near a stream of water.

*Habitat.* North and South Carolina, Massachusetts, Alabama.

[Specimens from Alabama are larger than those from New
England. *Supplement.*]

### 8. **Dolomedes sexpunctatus.**

Pl. 6, figs. 5, 6.

*Description.* Greenish; cephalothorax with a blackish margin, a white line each side, terminating at the anterior angle,
disc blackish-green, with a longitudinal paler line in the centre;
abdomen greenish-black, with four white dots near the base,
and four very minute ones nearer the apex. Male same color;
cephalothorax pale blackish-green, a pale yellowish line each
side; pectus pale, with six black dots; abdomen greenish-black
above, with four black rings near the base, sides and venter
[192] cinereous; trophi and first joints of feet pale testaceous
underneath; thighs unspotted apple green, the other joints
gradually deeper towards the tip.

*Observations.* This species dwells on ponds, and dives with
great agility, hiding itself under floating leaves or rubbish when
pursued.

*Habitat.* North Carolina.

[Pl. 18, fig. 55, eyes. Legs immaculate and hairy, arranged
4, 2, 1, 3. Taken February 28. *Supplement.*]

[♀, length 11.6 mm.; cephalothorax 5.4 mm.; legs 15, 15, 14.3, 16.7.
Abdomen whitish at the sides, and with two lines of white spots along

the back. The six spots under the thorax are indistinct, and in some specimens united into two brown bands.

Dorchester, Mass. April 27, on water.

Swampscott, " May 8, "

Dedham, " November 9, large number of young on a fence across a meadow.

Albany, N. Y. J. H. E.]

## Genus MICROMMATA. Latr. (*Sparassus*, Walck.)

Characters. *Chelicères moderately strong; maxillæ parallel, rounded at the extremity; lip short, rounded, wider near the base; eyes eight, subequal, in two rows, the upper one longest, curved from the base towards the lower row; feet long, slender, second pair longest, then the first and fourth, the third being the shortest.*

*Habits.* Araneides making no web for dwelling, but wandering and casting some irregular threads to arrest their prey; making a tent among leaves for the protection of the cocoon and the rearing of the young.

*Remarks.* I could not adopt the two families of Walckenaer, because the eyes, when large, are unequal; this shows the great affinity between Micrommata and Dolomedes. My first tribe, the Arcuatæ, approaches very closely to that subgenus, particularly *Micrommata undata.*

Tribe I. ARCUATÆ. *Lower row of eyes straight, middle eyes of the upper row larger, or borne on tubercles.*

Tribe II. BIARCUATÆ. *Upper and lower rows of eyes bent and opposed, the lower being bent towards the base, eyes equal or subequal.*

(*Arcuatæ.*)

### 1. Micrommata undata.

Pl. 6, fig. 7.

*Description.* Testaceous or yellowish; cephalothorax with [193] a broad, brownish band; abdomen with a scalloped, dusky band; feet slightly marked with dusky.

*Observations.* This spider is usually found on blossoms, watching for prey, in the manner of Thomisus. It seems perfectly distinct from *M. carolinensis*, but specimens occur ·in which the cephalothorax is much wider ; is it owing merely to the state of the abdomen, which has become narrower when the eggs are laid ? It makes no web, but, when attacked, it leaves a thread behind. This is a common species, which does not reach the size of *D. carolinensis.*

*Habitat.* Alabama. All seasons.

[Pl. 18, fig. 98, trophi. This species differs from *M. carolinensis* in its anterior eyes, which are in a straight line and by its cephalothorax, which has one broad band ; the abdomen has two hairy elevations anteriorly. *Supplement.*]

## 2. Micrommata serrata.

Pl. 6, fig. 8.

*Description.* Pale yellowish gray ; cephalothorax with two parallel, longitudinal, narrow, greenish bands ; abdomen with a narrow, scalloped, brownish band above, beneath with two longitudinal blackish lines, approaching each other, and becoming narrower towards the apex ; feet slightly marked with grayish rings, in all specimens, 2. 1. 4. 3. A small species.

*Observations.* This singular little spider is commonly found on plants, particularly on broad leaves, more abundantly in damp places. It spreads its feet, and seems fond of basking in the sunshine. Its webs are various ; sometimes it throws out only a few threads on the upper surface of a leaf ; at other times, it makes a web in the tops of bushes, like that of several species of Theridium ; and it has also been found in a web similar to that of Agelena, but open equally at both ends. I had once supposed that this might be the young of *M. undata* ; but I am convinced it is a very distinct species, never acquiring a large size.

*Habitat.* North Alabama. Summer.

[Pl. 18, fig. 1, eyes. Taken July 26. *Supplement.*]

44

(*Biarcuatæ.*)

### 3. Micrommata marmorata.

Pl. 7, fig. 5.

*Description.* Pale gray or whitish ; abdomen with an obsolete [194] scalloped band, grayish black ; feet, varied with
grayish black, 2. 1. 4. 3., or sometimes 2. 4. 1. 3. A large
species.

*Observations.* This spider lives on trees and bushes, where
it watches for prey, with extended legs. A female was found
in May, in the leaf of a *Morus multicaulis.* It had made its
cocoon there, and surrounded itself with a snow-white tent in
all directions. Transferred under a tumbler, it moved its
cocoon twice before it could be satisfied with a new location,
and made another smooth, white web. It remained constantly
by its cocoon, which it embraced closely with its long legs.
The cocoon is white, orbicular, and suspended by one thread in
the middle of the tent.

*Habitat.* North Alabama.

[Pl. 18, fig. 56, eyes ; fig. 105, trophi. Prof. Hentz had
formerly considered this to be the type of a new subgenus, for
which he gave the name of Dapanus, distinguished by having
its second pair of legs longest, the eyes subequal, the hinder
row curved posteriorly. *Supplement.*]

[Marietta, Ohio. Wm. Holden. J. H. E.]

### 4. Micrommata carolinensis.

Pl. 6, fig. 9.

*Description.* Testaceous or brownish ; cephalothorax with
two approximate longitudinal darkish bands on the disc ; abdomen with two longitudinal rows of abbreviated lines, and two
rows of small dots within these, white ; sometimes attaining
great size, 1.84, nearly two inches, from the end of the first
pair of legs to that of the fourth.

*Observations.* This spider is found wandering on trees,
walls, etc., and sometimes in houses, in search of prey. It is

very destructive of flies, and very voracious. Its cocoon, usually made under some large leaves, is white, orbicular. The mother hatches her progeny, and continues with the young for some time after they are come out ; the young living together under a common tent. A specimen was found in December, the cephalothorax of which was wider, the lower row of eyes straight, the abdomen small and tapering, with only a few abbreviated lines. Was this a distinct species ? A male was found, Alabama, April, measuring over three inches from the end of one of the second pair of legs to the end of the other.

*Habitat.* North Carolina, Alabama. [**195.**]

[The mandibles are very hairy at the top, 3 toothed ; feet arranged 2. 1. 4. 3. Taken from April to December. *Supplement.*]

[♀, length 13.4 mm.; cephalothorax 5.7 mm.; legs 23.5, 23,5, 20, 24.
♂ " 11 mm. " 4.6 mm. " 29, 29, 23, 26.2,
Young (*M. serrata*), length 3.8 ; legs 7, 7, 6, 7.
Pl. 20, fig. 4, adult ♀ ♂ 1 ♀ and ♀ palpus.
Hentz's drawings of *M. undata, serrata* and *carolinensis*, all represent young specimens, probably of one species, the adult form of which is described, but not figured, under the name of *M. carolinensis*. *M. undata* is the most common form. *M. serrata* is the young after the second moult.
West Roxbury, Mass. June 6, ♀ F. G. Sanborn.
Holyoke, Mass. July 4, ♀ and young in nests in grass.
Peabody, Mass. August 2, ♀ and young in nests in cedar tree.
Topsfield, Mass. September 3, ♀ and two hundred young in nests on low bushes.
Beverly, Mass. October 6, young ♀. J. H. E.]

## Genus OXYOPES. Latr. (*Sphasus*, Walck.)

Characters. *Cheliceres elongated ; fang short ; maxillæ narrow, elongated, tapering towards the tip ; lip as long as, or longer than the maxillæ, tapering towards the tip ; eyes eight, subequal or unequal, in four rows, two smallest ones forming the first, two largest ones forming the next, which is wider, two smaller ones forming the next which is widest, two small ones forming the last which is not as narrow as the first; feet, first pair longest, the second and fourth nearly equal, the third being shortest.*

*Habits.* Araneides wandering after prey, making no web, except around the cocoon, but casting some threads to secure their prey; cocoon conical, surrounded with points, placed in a tent made between leaves drawn together as a covering.

*Remarks.* The habits of this singular subgenus are very similar to those of the tribe Araboreæ, of the subgenus Dolomedes. They are found on the stems of trees, or on the blossoms of umbelliferous plants, with their legs extended, like Thomisus or Micrommata, and patiently waiting till some unsuspecting insect comes within their reach.

### 1. Oxyopes viridans.

Pl. 7, fig. 2.

*Description.* Tender grass-green; cephalothorax with small brown spots on the sides and at base; abdomen with yellowish, oval spots, edged with brownish, obliquely turned towards the centre, about four each side; feet and palpi pale, hairy; thighs and palpi with minute black dots beneath; feet, 1. 2. 4. 3. Large size.

*Observations.* This elegant species is by no means common. It is usually found on umbelliferous plants, where, like a Micrommata or Thomisus, it watches for the insects attracted by the blossoms. A specimen, taken in September, was kept several weeks in a glass vessel, where it soon made a cocoon [196] of a conical form, with small eminences, to which are attached the threads that hold it suspended firmly in the air, as that of *Theridium verecundum.* After it was finished, the mother watched it constantly, never leaving its unprotected family. Unfortunately, a rat, finding its way into the room, ate the watchful parent, leaving the cocoon, out of which the young were hatched on the 14th of October. These were of a deep orange color, measuring full 0.9 inch. The cocoon was of a pale greenish color. These habits show an affinity to Micrommata. It is possible that the mother carries its young like Lycosa.

*Habitat.* North Carolina, Alabama.

[Pl. 18, fig. 14, eyes; 14*a*, specimen from North Carolina; 14*b*, from Alabama; Pl. 19, fig. 134, cocoon. The difference in the eyes of these specimens is only apparent, and due to the quantity of hairs which lies across them. Taken September 1. *Supplement.*]

[♀, length 18 mm.; cephalothorax 6.8; legs 34.4, 30.2, 24, 26.5. Fernandina, Florida. E. Palmer. J. H. E.]

### 2. Oxyopes scalaris.

Pl. 7, fig. 4.

*Description.* Gray, varied with white and black; feet hairy.

*Observations.* This spider is usually found on trees, wandering after prey. It has the habitus of a Lycosa, and was observed by the writer for some time, before it was discovered that it belongs to a distinct subgenus. It is sought after by, and becomes the victim of the different species of the genus Sphex, a hymenopterous insect, which makes tubes of clay for the reception of its eggs, and in which it deposits great numbers of spiders, that are benumbed by its sting, but not entirely deprived of vitality, so that they continue alive till, the egg of the Sphex being hatched, the young larva finds in them fresh nourishment. It is common, where found at all.

*Habitat.* North Carolina.

[Pl. 19, fig. 120, the abdomen as it appears when empty. Pl. 7, fig. 4, represents it as it appears when full of eggs; feet arranged 1. 2. 4. 3. Taken in June. *Supplement.*]

### 3. Oxyopes salticus.

Pl. 6, fig. 10.

*Description.* Pale or yellowish; cephalothorax with four longitudinal blackish lines; abdomen, above, with various [197] slender, abbreviated, black and brownish lines, underneath whitish, with a longitudinal blank band; feet with very long hairs or bristles.

*Observations.* This species, usually found in the woods, is extremely active, leaping like an Attus. It is rather rare, but very distinct from any other.

*Habitat.* North Carolina, Alabama.

[Pl. 18, fig. 90, trophi. It leaps with more force and vivacity than an Attus. Taken in North Carolina in June. *Supplement.*]

### 4. Oxyopes astutus.

Pl. 7, fig. 1.

*Description.* Cephalothorax golden yellow, with four obscure, subobsolete, narrow bands; pectus with blackish marks each side; eyes, palpi, and trophi at base, black; abdomen greenish, with a silvery tinge; feet grassy green, hairy, 1. 2. 4. 3. or 1. 4. 2. 3.

*Observations.* This being adult, is much too small to be taken as the male of *O. viridans.* It has some affinity with *O. salticus,* but it is not probable that it will prove to be the male of that spider. When enclosed in a glass tube, it spun a web like that of Theridium, but composed of only a few threads. It was found in April, and also in September, strictly agreeing in size and markings.

*Habitat.* Alabama. September.

Genus LYSSOMANES. Mihi.

Characters. *Chelicores moderately strong; maxillæ parallel, short, rounded; lip conical, slightly truncated at tip; eyes eight, unequal, in four rows, the first composed of two very large eyes, the second of two smaller ones, placed farther apart, on a common elevation with the two forming the third row, which is narrower, the fourth about as wide, composed of two eyes placed on*

*separate elevations: feet, first pair longest, then the second, then
the third, the fourth being the shortest.*

*Habits.* Araneides wandering after prey, making no web,
cocoon. [198.]

*Remarks.* The singular spider which serves as the type of
this new subgenus, could not with propriety remain in the
subgenus Attus, in which the position of the eyes is subject to
very slight variations. Its habits are analogous. This is the
only spider in which the legs diminish in length from the first
pair to the fourth.

This subdivision will serve as link between Oxyopes and
Attus.

**Lyssomanes viridis.**

Pl. 7, fig. 3.

*Description.* Tender grass-green ; cephalothorax with some
orange-colored hairs near the eyes, and a little black line on its
disk; abdomen with six or eight black dots, sometimes wanting.
The two lowest large eyes are black, but appear green when
seen sideways ; the other six eyes stand on four tubercles.
Feet hairy, except the thighs, which are bare. 1. $\overline{2}$. 3. 4.

*Observations.* This elegant species is very active, and apparently fearless, jumping on the hand that threatens it.

*Habitat.* North and South Carolina.

[Pl. 18, fig. 91, trophi, wanting the palpus. Taken in April
and June. *Supplement.*]

Genus ATTUS. Walck. (*Salticus*, Latr.)

Characters. *Cheliceres strong, not long, except in some males ;
maxillæ parallel, widening above the insertion of the palpi, cut
obliquely above the lip ; lip as long as, or longer than, half the
length of the maxillæ, widest above the base, bluntly truncated at
tip ; eyes eight, unequal, in three rows, the first composed of four*

*eyes, the two middle ones largest, the second composed of two very small eyes, placed behind the external ones of the first, the third composed of two larger eyes, placed parallel to the second row; feet varying in length.* [**199.**]

*Habits.* Araneides wandering after prey, making no web, but concealing themselves in a silken valve, for the purpose of casting their skin, or for hibernation.

*Remarks.* I have formerly stated my reason for preserving the name Attus, given by Walckenäer to these Araneides. The species being very numerous, it would facilitate their study to arrange them in suitable subdivisions; but this is a difficult task. The families proposed by Walckenäer are vaguely characterized and insufficient. The relative position of the eyes offers some variations, but I could not succeed in obtaining satisfactory characters for subdivision from those variations. As the least objectionable mode, I have taken the relative lengths of the legs for the formation of my six families; that classification is somewhat artificial, but so is any other proposed. Moreover, the fifth tribe (that of the Saltatoriæ) offers a very natural subdivision. The third pair of legs, when longest, enables spiders to leap to an astonishing distance. The habits of the subgenus Attus will be best described by the history of the different species.

Tribe I. Pugnatoriæ, *first pair of legs longest and largest, the fourth next.*

#### 1. Attus audax.

Pl. 7, figs. 6, 7.

*Description.* Black; abdomen with a spot, several dots and lines, white; cheliceres brassy green; feet with gray and white hairs, 1. 4. 2. 3.

*Observations.* There is some obscurity in regard to the distinction between this and *A. 3-punctatus,* but there can be little doubt that there are two different species. This spider is very bold, often jumping on the hand which threatens it.

*Habitat.* Massachusetts. **[200.]**

[Northampton, Mass. Taken in May and July. *Supplement.*]

### 2. Attus insolens.

Pl. 7, fig. 8.

*Description.* Deep black; abdomen above, orange-red, with six blackish spots, wholly black beneath; cheliceres metallic green; the tip of the second joint of the palpi and the feet are varied with tufts of white hairs; the rest of the hair is black, except on the abdomen, where it is rufous above; feet, 1. 4. 2. 3. A male.

*Observations.* This species is probably rare, having occurred only once.

*Habitat.* North Carolina.

[Besides the second joint of the palpi and the feet, the knee of the first pair of legs is also varied with spots of white hairs. *Supplement.*]

### 3. Attus cardinalis.

Pl. 7, fig. 9.

*Description.* Scarlet; cephalothorax darker at base; cheliceres scarlet at base, steel-blue at their apex; palpi black; feet black, two last joints rufous at base, 1. 4. 2. 3.

*Observations.* I do not remember whether this spider was found by me, or given by Mr. Dutton.

*Habitat.* Southern States?

### 4. Attus capitatus.

Pl. 7, fig. 15.

*Description.* Piceous; cephalothorax with a narrow white band each side, and a whitish spot on the disc; second joint of palpi covered with white hairs; abdomen above with a

narrow, curved, yellowish white band near the sides, beneath
yellowish on both sides ; feet with a few white hairs, 1. 4. 2. 3.
A male.

*Observations.* This spider has great affinity with *Attus
militaris*, but is sufficiently distinct. The female probably
differs from this in markings, and possibly is among my [201]
descriptions ; but this can be established only by future observers, who, after all my labors, have still a wide field before
them to perfect the history of the spiders of North America.
This was communicated to me by Mr. Thomas R. Dutton,
a young naturalist of great perseverance, energy, and discrim-
ination, who collected it in Georgia.

[Pl. 18, fig. 26, eyes. The mandibles have not so sharp an
inner point as *A. militaris* ; the white band on the cephalothorax reaches neither the base nor the front ; the yellowish white
band on each side of the abdomen is blackish on the extreme
sides ; in the description it is stated that the second joint of the
palpi is covered with white hairs ; on the sheet containing the
drawing it is stated that it is the first joint which is so characterized. *Supplement.*]

### 5. Attus militaris.

Pl. 7, figs. 10, 11.

*Description.* Rufous, varied with brown ; cephalothorax
with one, sometimes two, white spots ; abdomen above with
two longitudinal blackish bands, on which are oblong white
dots, which near the base are usually joined so as to form a
narrow band, beneath whitish with a blackish longitudinal
band. Male rufous or piceous ; cephalothorax with a spot and
a band around the anterior portion, and a narrow longitudinal
line on the disc, white ; abdomen above with a white band
on the margin, which does not quite reach the apex, pale
grayish brown beneath ; feet, in the female, 1. 4. 3. 2., in the
male, 1. 4. 2. 3.

*Observations.* Much as the sexes differ from each other, I cannot doubt their constituting one species, having repeatedly found them enclosed quietly in the same silk tube, and having always found the males and the females with the characters given above. The spots and markings of these spiders are formed by hairs or scales, which have certain metallic reflections. The motions of this spider are slow, and exhibit caution; it is found usually on trees, and often hibernates under the bark of decaying trunks. The male, remarkable for his enlarged, nearly horizontal cheliceres, is a very bold little fellow, always ready for action, and determined to see all things for himself, raising and turning his head towards the object that approaches him, and usually jumping upon his enemy instead of ingloriously retreating. This species is a common one.

*Habitat.* North Carolina, Alabama. [202.]

[Taken in March, May and December. ♂ with the abdomen covered on the disc with golden hairs or scales; the legs also with more scattered hairs of the same color. *Supplement.*]

[Mayport, Fla.; Marietta, Ohio, ♂; Charlestown, Mass., ♂; Hyde Park, Mass. Wm. Holden. J. H. E.]

### 6. Attus multicolor.

Pl. 7, fig. 13.

*Description.* Cephalothorax black, with a pale, irregular band each side of the disc; abdomen metallic green, with a band at base, and a diagonal spot each side, orange, and with eight small white spots; underneath obscure gray, with inflections of green on the pectus; feet rufous or pale, varied with piceous, 1. 4. 2. 3.

*Observations.* This species is related to *A. otiosus* and *mystaceus,* but distinct from both by the absence of the tufts of hair on the cephalothorax, and other characters. The palpi are pale yellow, and there is a black band more or less visible on each side of the abdomen.

*Habitat.* Alabama. June – August.

### 7. Attus sexpunctatus.

Pl. 7, fig. 14.

*Description.* Black; cephalothorax with the two posterior eyes near the base, which is wide and suddenly inclined at nearly a right angle with the upper surface, cheliceres with a strong inner tooth, and a long, curved fang; abdomen with six dots, and a line in front, white; feet, 1. 4. 2. 3., first pair with enlarged thighs and quite long.

*Observations.* This cannot be confounded with *Attus fasciolatus*, which is also designed from a female. By the characters derived from its cheliceres, it approaches Epiblemum. I suppose it must be a rare species, having never met with any other specimen.

*Habitat.* North Carolina.

[Taken in July. *Supplement.*]

———

[Continued from Vol. v, p. 202.]

### 8. Attus falcarius.

Pl. 8, fig. 1.

*Description.* Cephalothorax and abdomen covered with yellowish gray hairs, hairs longer in front of the abdomen; feet, 1. 4., very stout, 2. 3.

*Observations.* Very distinct from any other by the form of its abdomen.

*Habitat.* Alabama.

[Pl. 18, fig. 35, eyes. Taken August 6. *Supplement.*]

### 9. Attus binus.

Pl. 8, fig. 2.

*Description.* Blackish; abdomen pale bluish gray; with two parallel, longitudinal, blackish lines above; feet, 1. 4. 3. 2.

*Observations.* I never found more than one specimen of this very distinct species. Its abdomen was very much distended, and it moved very slowly.

*Habitat.* Found on Sullivan's Island, South Carolina.

Tribe II. LUCTATORIÆ. *Fourth pair of legs longest, the first next and largest.*

### 10. Attus Nuttallii.

Pl. 8, fig. 3.

*Description.* Piceous; abdomen pale gray above, with an oblong scalloped, black, longitudinal band surrounding a small white spot; feet, 4. 1. 2. 3.

*Observations.* This probably very rare species was found in the hot-house of the botanic garden at Cambridge, in the [353] presence of the distinguished botanist and ornithologist, Thomas Nuttall.

*Habitat.* Massachusetts.

### 11. Attus castaneus.

Pl. 8, fig. 4.

*Description.* Black or piceous, with some long black hairs, and short, thick, yellowish down, particularly distinct on the abdomen, which has a whitish line at base, continued on the sides to near the middle; sides of the abdomen, with oblique lines, whitish; venter with four white lines, all the lines being formed by whitish hairs; dorsum with four or six obsolete dots; feet rufous, with blackish rings, 4. 1. 2. 3., the fourth longest and slender, the first next, very stout.

*Observations.* This spider is perfectly distinct from any other yet observed. It must be rare, having occurred only once, under a stone, in March.

*Habitat.* North Carolina.

[Pl. 18, fig. 36, eyes. *Supplement.*]

## 12. Attus tæniola.

Pl. 8, fig. 5.

*Description.* Black ; cephalothorax with a white fillet on each side, continued to near the base ; abdomen with two longitudinal, narrow lines, composed of white dots or abbreviated lines ; tarsi dark rufous or blackish. 4. 1. 2. 3.

*Observations.* This is not a rare species, and shows only a moderate degree of activity.

*Habitat.* North Carolina, Alabama.

[Taken in May. *Supplement.*]

## 13. Attus elegans.

Pl. 8, fig. 6.

*Description.* Pale rufous ; cephalothorax with eyes nearer the apex than the base, second joint of palpi piceous ; abdomen [354] metallic green with yellow and red reflections, a white band, widest in front and continued on the sides, but not reaching the eyes ; feet, 4. 1. 3. 2., with a slender black edge externally, thighs of first pair black, knee pale.

*Observations.* This graceful species is readily distinguished from any other, and is not very rare.

*Habitat.* Southern States.

[Pl. 18, fig. 2, eyes. Taken in July. *Supplement.*]

Tribe III. INSIDIOSÆ. *Legs equal in thickness, the fourth longest, then the first.*

## 14. Attus familiaris.

Pl. 8, fig. 7.

*Description.* Pale gray, hairy ; abdomen blackish, with a grayish, angular band, edged with whitish ; feet, 4. 1. 2. 3.

*Observations.* This very common spider, almost domesticated in our houses, by its habits, deserves a longer notice than others. It dwells in cracks around sashes, doors, between clapboards, etc., and may be seen on the sunny side of the house, and in the hottest places, wandering in search of prey. It moves with agility and ease, but usually with a certain leaping gait. The moment, however, it has discovered a fly, all its motions are altered ; its cephalothorax, if the fly moves, turns to it, with the firm glance of an animal which can turn its head ; it follows all the motions of its prey with the watchfulness of the falcon, hurrying its steps or slackening its pace, as the case may require. Gradually, as it draws near to the unsuspecting victim, its motions become more composed, until, when very near, its movements are entirely imperceptible to the closest observation, and, indeed, it would appear perfectly motionless, were it not for the fact that it gradually draws nearer to the insect. When sufficiently near, it very suddenly takes a leap, very seldom missing its aim. I saw one, however, make a mistake, for the object which it watched was only a portion of the wing of an hemipterous insect entangled in a loose web. It took its leap and grasped [355] the wing, but relinquished it immediately, apparently very much ashamed of having made such a blunder. This proves that the sight of spiders, though acute, is not unerring. Before leaping, this Attus always fixes a thread on the point from which it departs; by this it is suspended in the air if it miss its aim, and it is secure against falling far from its hunting grounds.

These spiders, and probably all other species, a day or two before they change their skin, make a tube of white silk, open at both ends; there they remain motionless till the moulting time arrives, and, even some days after, are seen there still, probably remaining in a secure place, for the purpose of regaining strength and activity.

*Habitat.* Throughout the United States.

[Pl. 18, fig. 74, eyes ; fig. 90, trophi. *Supplement.* ]

[♀, length 10.5 mm.; cephalothorax 4.4 mm.; legs 9, 7.8, 7.8, 9.
♂     "     10 mm.         "       4.6 mm.;  "  11.5, 9.3 8. 9.8.

Palpus of ♂. Pl. 20, fig. 5.

Salem, Mass. May 30, ♂ and ♀ on fences. October 27, in bags under bark.

Providence, R. I. November 11, in bags under bark.

Ohio, ♂ ♀, Northern Illinois, ♀, Wm. Holden. J. II. E.]

## 15. Attus tripunctatus.

Pl. 8, fig. 8.

*Description.* Black; abdomen, with metallic reflections and white and orange-colored hairs, with a central spot and two short bands white, which are surrounded with deep black; cheliceres brassy green; feet, $\overline{4. 1. 3. 2.}$

*Observations.* This is perhaps the most common Attus in the United States. It is usually found on dead trees, under the bark of which it takes refuge, and also hibernates there, in tubes of strong white silk. The spots are often of an orange color, instead of being white.

*Habitat.* The United States.

[Pl. 18, fig. 75, eyes and extremities of cheliceres ; fig. 106, trophi. Very common in New England. *Supplement.*]

[♀, length 8.6 mm. ; cephalothorax 4 mm.; legs 8.2, 6.5, 6.5, 8,5.

♂ " 8 mm. " 3.5 mm.; legs 8.2, 6.5, 6, 7.2.

Palpus of ♂. Pl. 20, fig. 6.

Salem, Mass. March, in bags under stones. May 20, ♂.

Beverly " July 1, young.

Boston, " December 7, in tubes in cocoons of *Epeira riparia.*

Providence, R. I. Indianapolis, Indiana.

Ohio, ♂ ♀ ; Ann Arbor, Mich., ♂ ♀ ; Ft. Towson, Red River, Ark., ♂ ♀ ; Knoxville, Tenn., ♀ ♂ ; East Florida, ♀ ; North Carolina, ♀. Wm. Holden. J. II. E.]

## 16. Attus mystaceus.

Pl. 8, fig. 9.

*Description.* Gray ; varied with whitish spots ; cephalothorax with four tufts of bristles in the region of the eyes ; feet, 4. 1. 2. 3.

*Observations.* This large and very distinct species is not [356] rare on the eastern side of the Alleghany mountains, as far north as the 35° of latitude; but it has not been found by me in Alabama.

*Habitat.* North Carolina.

[Pl. 18, fig. 76, eyes. Pl. 19, fig. 119, lateral view. Specimens taken in the fall were kept through the winter. *Supplement.*]

[♀, length 11.6 mm.; cephalothorax 5.2 mm.; legs 9.6, 8, 5, 8, 10.6.

This is one of the most common species of Attus around Boston. It is found at all seasons in thick tubes of white silk, under stones. None of my specimens have the tufts of hairs on the head as distinct as in Hentz's figure. I do not know the male.

Salem, Mass. March, old and young in bags under stones. July 18, ♀, in bag with cocoon of young. September and October, in bags under stones.

Providence, R. I.

Ohio, ♂ ♀; St. Louis, Mo.; Ft. Towson, Red River, Ark., ♀. Wm. Holden. J. H. E.]

### 17. Attus otiosus.

Pl. 8, fig. 10.

*Description.* Blackish, mostly covered with white hairs; cephalothorax black at base and anteriorly, two tufts of hairs each side on the region of the eyes; abdomen with a band at base, and several angular spots, white, and with a longitudinal green band more or less covered with hairs and edged with a scalloped black line each side, beneath white with a black band very wide at base, and tapering towards the apex where it branches out; feet varied with rufous and black, 1. 4. 2. 3., the fourth slightly longest when separated from the body. A large species.

*Observations.* This spider, related to *A. mystaceus,* was found in mid-winter, enclosed in silk tubes, under the bark of dead trees, where great numbers were hibernating.

*Habitat.* North Alabama.

[The legs are varied with rufous and black, with tufts of whitish hairs; the spots on the body vary a little in different specimens. *Supplement.*]

### 18. Attus fasciolatus.

Pl. 8, fig. 11.

*Description.* Black; cephalothorax with three grayish spots; abdomen with three small spots, two abbreviated lateral lines, and an anterior one white; feet varied with rufous, 4. 1. 2. 3.

*Observations.* This spider seems to be quite distinct from. *A. tripunctatus*, but may prove only a variety of that species.

*Habitat.* South Carolina, Massachusetts.

[Pl. 18, fig. 63, eyes. *Supplement.*]

### 19. Attus rufus.

Pl. 8, fig. 12.

*Description.* Rufous; abdomen with a yellowish white [357] band anteriorly which extends to the sides, on the disk four white dots, and four smaller grayish ones, the dots surrounded by black rings which usually unite in the form of a longitudinal band on each side, beneath pale, with three subobsolete longitudinal lines; feet, 4. 1. 2. 3. or 3. 2., in the male 1. 4. 2. 3.

*Observations.* This spider, which is not very common, is found on plants, and is not remarkably active. In the male, the abdomen is white around and between the bands.

*Habitat.* United States.

[Pl. 18, fig. 37, eyes. Body covered with thick and long white hairs; cheliceres bright rufous, black at the apex, with a line of white hairs between them and the eyes. Alabama in May, July and August; in Carolina in August; and in Massachusetts in the collection of Prof. Peck, taken in July. *Supplement.*]

### 20. Attus podagrosus.

Pl. 8, fig. 13.

*Description.* Cephalothorax piceous black ; abdomen pale brownish, white at base, with a scalloped dusky band ; feet bright rufous, joints tipped with black, with some hairs, 4. 1. 2. 3. A large species.

*Observations.* This may be readily distinguished from *A. rupicola,* to which it is closely related.

*Habitat.* Alabama. November.

[Cheliceres darkish, but not green. *Supplement.*]

### 21. Attus rupicola.

Pl. 8, fig. 14.

*Description.* Rufous, very hairy; abdomen brownish, with a paler band and two blackish dots ; feet varied with blackish, in the female 4. 1. 2. 3., in the male 1. 4. 2. 3. A large species.

*Observations.* The male, which resembles the female, has invariably its first pair of legs longest and stoutest. This species was repeatedly found in cavities of limestone rocks on the margin of a river, moving cautiously and slowly on the surface of the stones, and retreating into crevices.

*Habitat.* Alabama. September. [358.]

### 22. Attus nubilus.

Pl. 8, fig. 15.

*Description.* Pale gray ; cephalothorax with a tinge of rufous at base, and many obscure markings : abdomen with obscure, waved bands ; feet with blackish rings, 4. 1. 2. 3. A somewhat small species.

*Observations.* This spider is common, usually found on the stems of plants.

*Habitat.* Alabama. May – July.

[Pl. 18, fig. 27, eyes. *Supplement.*]

**23. Attus hebes.**

Pl. 8, fig. 16.

*Description.* Brownish; abdomen white, with a greenish spot surrounded with four black dots, near the base, and a black fascia near the apex; feet, 4. 1. 3. 2.

*Observations.* This probably rare species was found on the ground, having fallen from a tree.

*Habitat.* Massachusetts.

[Pl. 18, fig. 3, eyes. Taken in June. *Supplement.*]

**24. Attus parvus.**

Pl. 8, fig. 17.

*Description.* Grayish; abdomen with six or eight abbreviated transverse lines, white; feet varied with . rufous and black, 4. 1. 2. 3.

*Observations.* A somewhat obscure species, which I believe I have seen in the North.

*Habitat.* North Carolina, Massachusetts?

**25. Attus rarus.**

Pl. 8, fig. 18.

*Description.* Blackish; cephalothorax with green scales, and some yellow ones anteriorly; abdomen with green scales, [359] except on a black band which surrounds the disk, a yellow band at base, extending each side, but which does not reach the middle, one large yellow dot each side near the middle, two little dots on the disk, and four terminal abbreviated bands white; beneath blackish, abdomen with some yellowish hair which forms two or four sub-obsolete, abbreviated, longitudinal lines; feet, 4. 1. 2. 3.

*Observations.* This very distinctly-marked species is probably very rare, as it occurred only once.

*Habitat.* North Carolina. June.

### 26. Attus niger.

Pl. 8, fig. 19.

*Description.* Deep black ; legs pale testaceous, 4. 1. $\overline{3.\ 2.}$

*Observations.* This small species is remarkable on account of its activity in running and leaping.

*Habitat.* North Carolina.

[Pl. 18. fig. 4, eyes. Taken in July. *Supplement.*]

### 27. Attus? gracilis.

Pl. 8, fig. 20.

*Description.* Rufous ; cephalothorax very prominent ante-, riorly, wider behind the middle ; abdomen narrower, slender, fusiform, nipples long ; feet long and slender, $\overline{4.\ 1.}\ \overline{3.\ 2.}$

*Observations.* This cannot be *Synemosyna scorpionia* ; but may ultimately be referred to that division.

*Habitat.* Alabama. August.

[Pl. 18, fig. 57, eyes ; fig. 107, trophi. *Supplement.*]

Tribe IV. METATORIÆ. *Legs subequal in thickness, the fourth longest, then the third.*

### 28. Attus leopardus.

Pl. 8, fig. 21.

*Description.* Cephalothorax black, rufous about the eyes, [360] with a curved white line each side ; abdomen with two opposed lenticular black bands surrounded with white, pale gray underneath, with two sub-obsolete longitudinal, whitish lines ; feet rufous with many black rings, 4. 3. $\overline{2.\ 1.}$

*Observations.* This spider is common. The female is often found under stones with its cocoon, which is white.

*Habitat.* Alabama. May.

[♀, length 9.2 mm.; cephalothorax 3.7 mm.; legs 4.5, 4.5, 5.8, 7.5.
One specimen from Mt. Tom, Holyoke, Mass.
Marietta, O., ♂ ♀ ; Rushville, O., ♀. Wm. Holden. J. H. E.]

### 29. Attus puerperus.

Pl. 8, fig. 22.

*Description.* Testaceous or yellowish ; intermediate small eyes, and the two last, borne on elevations ; abdomen with about twelve black dots, underneath with a black spot near the apex ; feet, 4. 3. or 3. 4. 1. 2.

*Observations.* Mr. Thomas R. Dutton, who brought this from Georgia, gave me another one, which was not, like this, replete with eggs. The abdomen not being distended, the dots appeared less regular and distinct.

*Habitat.* Georgia.

[Pl. 18, fig. 28, eyes. *Supplement.*]

### 30. Attus vittatus.

Pl. 8, fig. 23.

*Description.* Cephalothorax and trophi rufous varied with blackish ; abdomen gray, with reddish curved bands ; feet pale rufous or yellowish, 4. 3. 1. 2., in the male 4. 1. 2. 3., and speckled with black dots.

*Observations.* With some hesitation I refer to the same species the drawings of a male, and that of a female, which I had considered as distinct, on account of the difference in the respective length of the legs. In the genus Attus that character is sometimes a sexual distinction.

*Habitat.* North Carolina, Alabama. [361.]

[Cephalothorax with two rufous conic spots united at base ; body pale beneath. Taken in May. *Supplement.*]

[Malden, Mass., ♀. Wm. Holden. J. H. E.]

Tribe V. SALTATORIÆ. *Third pair of legs longest, then commonly the fourth.*

### 31. Attus coronatus.

Pl. 9, fig. 1.

*Description.* Pale dusky; cephalothorax varied with black, a scarlet spot between the eyes and the cheliceres ; abdomen with two curved bands and about three spots, white; pale beneath without distinct spots; legs, with first pair stoutest, black on the internal side, 3. 4. 1. 2.

*Observations.* The bright scarlet spot on its front gives to this spider a whimsical air of fierceness, which is heightened by its attitudes and singular motions. The lighter spots on the cephalothorax are produced by yellowish hairs. It is not very rare. It is probably quite distinct from *A. cœcatus.*

*Habitat.* Alabama. May – July.

[Pl. 18, fig. 82, eyes. *Supplement.*]

### 32. Attus cœcatus.

Pl. 9, fig. 2.

*Description.* Brownish obscure; cephalothorax with a red spot under the eyes, and with a basal spot and large fascia black; abdomen varied with black and brownish obscure, pale bronzed beneath; feet, first pair stoutest, black with a line of yellowish scales above, antepenultimate joint with two long, black scales or spatulæ, thighs with thick tufts of black hairs, the other legs varied with black and brownish, 3. 4. 1. 2. A small species.

*Observations.* This species, though very different in marking, is very closely related to *A. coronatus.*

*Habitat.* Alabama. September.

### 33. Attus pulex.

Pl. 9, fig. 3.

*Description.* Pale brownish; cephalothorax large, varied [362] with piceous, edged widely with blackish towards the base; abdomen nearly orbicular, piceous, varied with whitish spots, and a band at base; feet varied with piceous, 3. 4. 1. 2. A small species. Male like the female.

*Observations.* This little spider is common near the ground, where it may be seen moving with sudden, rapid motions, and

jumping, like a flea, to great distances. It is a well-character-
ized species.

*Habitat.* Alabama. April – May.

[A male one-third as large as the specimen figured, Pl. 9,
fig. 3, was taken May 29. *Supplement.*]

### 34. Attus roseus.

Pl. 9, fig. 4.

*Description.* Cephalothorax white, blackish at base; abdomen roseate, with a whitish base; feet pale yellow, 3. 4̄. 1̄. 2.

*Observations.* This small species is not unfrequently found
on grass, in May and June.

*Habitat.* Massachusetts.

[Pl. 18, fig. 15, eyes. *Supplement.*]

### 35. Attus viridipes.

Pl. 9, fig. 5.

*Description.* Cephalothorax rufous, with black bands and
spots; abdomen white, with two black angular bands; anterior
feet greenish; the other feet varied with rufous, blackish and
white, 3. 1. 4. 2.

*Observations.* This small spider is usually found on the
ground, on sand or on grass, in constant activity. When any
object approaches it, it lifts itself on its posterior limbs to
reconnoitre the enemy or the prey. It never was seen large.

*Habitat.* South Carolina.

[Pl. 18, fig. 64, eyes. *Supplement.*]

### 36. Attus auratus.

Pl. 9, fig. 6.

*Description.* Black; palpi, sides of the cephalothorax and four
spots above, silvery white; abdomen with a cross and circular
band, golden color; feet varied with rufous, 3̄. 4. 1. 2. [363.]

*Observations.* This beautiful species seems to fear the light ; for I never found it except when enclosed in the old shells of the pupæ of some hymenopterous insect. It is rare.

*Habitat.* South Carolina.

[Pl. 18, fig. 65, eyes ; fig. 92, trophi. *Supplement.*]

[Mayport, Fla., ♀. Wm. Holden. J. H. E.]

### 37. Attus multivagus.

Pl. 9, fig. 7.

*Description.* Piceous ; palpi pale ; abdomen gray, with curved bands, dots and a spot white, pale underneath with a longitudinal darkish line and a pale one each side, all sub-obsolete ; feet, 3. 4. 1. 2. A middle-sized species.

*Observations.* This species in markings resembles *A. fasciolatus,* but is quite distinct from it.

*Habitat.* Alabama. April.

[Zanesville, O., ♀. Wm. Holden. J. H. E.]

### 38. Attus cristatus.

Pl. 9, fig. 8.

*Description.* Pale brownish ; cephalothorax with small dusky marks, palpi very small ; abdomen with curved dusky lines, and a tuft of white hairs at base, pale underneath, with two sub-obsolete, approximate longitudinal paler lines ; feet pale, 3. 4. 1. 2.

*Observations.* The tuft of white hairs on the base of the abdomen, and projecting over the cephalothorax, is not peculiar to this species alone, but by other characters it is sufficiently distinguished.

*Habitat.* Alabama. July – August.

[Pl. 19, fig. 112, lower surface of abdomen. *Supplement.*]

Tribe VI. Ambulatoriæ; *legs usually slender, the first pair longest, the fourth next.*

### 39. Attus mitratus.

Pl. 9, fig. 9.

*Description.* Pale above and beneath; cephalothorax with [364] a broad pale brownish band; abdomen with a pale brownish band, interrupted with yellowish in about three places; feet, 1. 4. 2. 3. A small species.

*Observations.* This is not a rare species. It is usually found on plants, moving slowly on the stems.

*Habitat.* Alabama. April – May.

[Only males were found. *Supplement.*]

### 40. Attus sylvanus.

Pl. 9, fig. 10.

*Description.* Piceous; cephalothorax reddish anteriorly, with a yellowish spot on the disk, and four oblique slender lines of the same color; abdomen with two parallel longitudinal yellowish lines; thighs rufous at base, except the first pair; feet, 1. 3̄. 2̄. 4̄.

*Observations.* This graceful species is found commonly on the trunks of trees, moving rather slowly, and walking backwards when threatened by an enemy. It moves its anterior feet like palpi, as if to feel its way in its progression.

*Habitat.* South Carolina.

[Pl. 18, fig. 58, eyes; fig. 108, trophi. *Supplement.*]

### 41. Attus superciliosus.

Pl. 9, fig. 11.

*Description.* Cephalothorax black between the eyes, deep ferruginous at base, covered anteriorly with golden or greenish scales, a tuft of hairs between the eyes; abdomen black, with the same kind of scales, the absence of which forms obsolete blackish lines on the disk, beneath with such scales also; pectus and thighs glabrous, ferruginous; feet with a black fillet externally, antepenultimate joint of first pair with a tuft of black hairs, 1. 4. 2. 3.

*Observations.* This singular species can be readily distinguished by the tuft of hairs placed above the lower row of eyes, and resembling eyebrows. It is probably rare.

*Habitat.* North Carolina. [365.]

[Pl. 18, fig. 5, eyes. On the antepenultimate joint of all the legs there is a black fillet on the anterior side, which is faintly continued on the preceding and following joints, and even on the thighs. Taken in June. *Supplement.*]

### 42. Attus morigerus.

Pl. 9, fig. 12.

*Description.* Cephalothorax ferruginous, covered with silvery down, through which the color can be seen, particularly about the eyes; abdomen above dark brown, covered with silvery down, four spots and a band glabrous; beneath pale : feet pale yellowish, with some hairs, 1. 4. 2. 3.

*Observations.* This little spider may be seen usually on leaves, where it frequently makes its tubes. It has been seen on the hickory and the mulberry trees.

*Habitat.* North Carolina, Alabama. April, May.

[Taken October 17. *Supplement.*]

### 43. Attus cyaneus.

Pl. 9, fig. 13.

*Description.* Brassy green ; body short; feet, 1. 4. 3. 2. Small.

*Observations.* This small but brilliant spider is found on plants, during all the warm season.

*Habitat.* North Carolina, Alabama.

[Pl. 18, fig. 66, eyes. Taken in April, May, June, etc. *Supplement.*]

## 44. Attus canonicus.

Pl. 9, fig. 14.

*Description.* Rufous, or deep orange; abdomen with a longitudinal row of black dots, seven or eight on each side above; feet with black rings; cephalothorax and anterior part of the abdomen covered with dense yellowish rufous hair. Feet, 1. $\overline{4}$. $\overline{2. 3}$.

*Observations.* Found in Cambridge, Massachusetts, in August.

*Habitat.* Massachusetts.

[Pl. 18, fig. 6, eyes. The jaws are very short. *Supplement.*]

## 45. Attus octavus.

Pl. 9, fig. 15.

*Description.* Grayish brown; abdomen above with eight [366] large black dots, two green spots, and some white marks, gray beneath; feet rufous, 1. 4. 2. 3.

*Observations.* This is a common species in the south. A specimen was found with legs 4. 1. 3. 2., shorter, and with blackish rings. Is it a different species? It is not probable that this can be referred to *A. hebes.*

*Habitat.* Alabama. July – August.

Genus EPIBLEMUM. Mihi.

*Characters. Cheliceres very long, slender, horizontal, in both sexes, fang nearly as long; maxillæ parallel, wide at base, narrowed above the insertion of the palpi, cut obliquely on both sides towards the point; lip conical; eyes eight, unequal, in three rows, the first composed of four, the two middle ones somewhat larger, the second composed of two very small ones placed nearer the third row, which is composed of two larger ones; feet, first pair longest, then the fourth, the third or second shortest.*

*Habits. Araneides wandering after prey, making no web, cocoon.*

*Remarks.* The characters of this subgenus are quite sufficient to separate and distinguish the species composing it from Attus. Even allowing that the character derived from the extreme length of the cheliceres were limited to the males, the great number of species contained in Attus would authorize naturalists to separate such as have that character under a separate denomination. But it seems that this peculiarity may be confined to the females in some species ; as, a male of *E. palmarum* was found with short cheliceres ; but these were nevertheless horizontal.

## 1. Epiblemum palmarum.

Pl. 9, fig. 16.

*Description.* Rufous or dark brown ; cephalothorax and [367] abdomen with a whitish band on each side above ; feet whitish, except the first pair which are rufous, 1. 4. 2. 3.

*Observations.* Cuvier, in his Règne Animal, IV, p. 264, says that some males of Attus have elongated cheliceres. But this was a female ; and a male was found in North Carolina, corresponding to this in every particular, except that the cheliceres were not elongated, but *they were horizontal.* The subgenus Attus is so large that some good subdivision is required. Like Tetragnatha, this spider extends its legs in one line along the twig or blade on which it rests.

*Habitat.* South and North Carolina.

A male was found in Alabama, corresponding with this in every respect. He was bold, and moved with a ludicrous motion of his first pair of legs, which he waved to and fro, in advancing towards the body which was extended against him.

## 2. Epiblemum faustum.

Pl. 9, fig. 17.

*Description.* Piceous ; cephalothorax with the margin and two spots white ; abdomen with the base and four short lines white ; feet, 1. 4. 3. 2.

*Observations.* This species was found common in Cambridge, Massachusetts, on walls, on the south side.

*Habitat.* Massachusetts.

[Pl. 18, fig. 59, eyes; fig. 109, trophi. Taken in June. *Supplement.*]

[Pl. 20, fig. 8. Adult ♂ and ♀, and palpus of ♀.

The females do not have long horizontal mandibles like the males. Probably identical with *Salticus scenicus* Blackw., Spiders of Great Britain and Ireland, and *Callietherus histrionicus* Sim., Monograph. des Attides, Ann. Soc. Ent. France.

Salem, Mass., Providence, R. I., Albany, N. Y., on fences and houses at all seasons.  J. H. E.]

## Genus SYNEMOSYNA. Mihi.

*Characters.* *Chelicerce short in the females; maxillæ slightly inclined toward the tip, truncated at tip; lip short, rounded; eyes eight, unequal, in three rows, the first composed of four eyes, the two middle ones largest, the second composed of two small ones placed nearer the first than the third, which is composed of two larger eyes; feet slender, the fourth pair longest, the other three variable; body elongated, nodose, abdomen contracted near the middle.*

*Habits.* Araneides wandering after prey, making no web, but silk tubes, for hibernation, running on plants like ants, which they resemble; cocoon.

*Remarks.* This differs in many points from Myrmecia, Latr., Ann. des Sc. Nat. IV. p. 261, and yet seems to be closely related to it. That subdivision is not known to me, though it is said in that work that some species are found in Georgia. In Myrmecia the chelicercs are large, in this, they are small, at least in the females; in that subgenus the maxillæ are rounded and hairy, the abdomen is much shorter than the cephalothorax, and they have other characters which do not belong to this.

I have already pointed out the features, and proposed a name for this singular subdivision, in a paper published in Silliman's Journal. I have, since writing that article, discovered one species, in addition to the three mentioned there. They are all anomalous, and differ from each other in many points; while they agree in the characters which I have assigned. They hibernate in silk tubes, under the bark of trees.

### 1. Synemosyna formica.

Pl. 9, fig. 18.

*Description.* Rufous; cephalothorax very long, contracted in the middle, tapering towards the base, and with two lateral yellowish spots; abdomen contracted in the middle, also with two lateral yellow spots, each where the contraction appears; feet slender, varied with yellowish and black, 4. 3. 1. 2., tibiæ of the first pair and part of the tarsus black underneath. Male with very large cheliceres; legs, 4. 1. 3. 2.

*Observations.* This spider cannot be placed in the subgenus Myrmecia, of Latreille, as described in the fourth volume of the Ann. des Sc. Nat., or in Vol. iv. p. 261 of the Règne Animal, [369] for the following reasons; the eyes are very unequal in size, and not placed in the manner described; the cheliceres are large only in the males; and the length of the feet is not the same. It is possible, however, that the insects drawn by Abbot belong to this division; for, being very small, probably the situation of the eyes may not have been correctly observed. Be this as it may, the subgenus Myrmecia, or Myrmecium, is closely related to this.

I had seen individuals of this species running on the blades of grass and stems of weeds, long before I distinguished them from ants. They move with agility and can leap, but their habitus is totally different from Attus. They move by a regular progression or regular walk, very different from the halting gait of that subgenus.

*Habitat.* North Carolina, Alabama.

74

[Taken in April, May and July. *Supplement.*]

[Length of ♀ 4 mm.; legs 2.2, 2.4, 3.6.
Palpus of ♂. Pl. 20, fig, 9.
The only male 1 have seen has short mandibles, and is otherwise much
like the female.
Ipswich, Mass.    June 17, ♂, F. G. Sanborn.
Holyoke,   "      July 4, many females.
Malden,   "      H. L. Moody.
Washington, D. C.   E. P. Austin.   J. H. E.]

## 2. Synemosyna scorpionia.

Pl. 9, fig. 19.

*Description.*  Piceous; cephalothorax with two sub-obsolete,
pale spots; posterior eyes placed near the base, and remote
from the rest; abdomen slightly contracted near the middle,
with a yellowish indented spot; feet rufous, 4. 1. 2. 3., first
pair very stout; sexes alike, the cheliceres not being enlarged
in the male.

*Observations.*  This small spider is somewhat rare, and was
found in the winter months.

*Habitat.*  North Carolina.

[Pl. 18, fig. 67, eyes.  The ♂ was taken in November; the
♀ in February, a little larger than the ♂, and with the abdomen very slightly contracted.  *Supplement.*]

[Marietta, Ohio.  Wm. Holden.  J. H. E.]

## 3. Synemosyna ephippiata.

Pl. 9, fig. 20.

*Description.*  Rufous; cephalothorax wide in the region of
the eyes, tapering towards the base; abdomen depressed before
the middle, widest beyond the middle, a transverse paler band
near the middle, piceous towards the apex; feet, [370] with
the interior edge black, last two joints of second pair black,
penultimate and antepenultimate joints of the leg of the fourth
pair dusky, 4. 2. 3. 1.

*Observations.* This is a very distinct species, found hibernating in silk tubes under bark, making such tubes when confined. The male, with cheliceres not enlarged, was found agreeing with the above description in the minutest particular. This shows beyond any doubt that the species is distinct from *S. formica.*

*Habitat.* Alabama. December.

[Pl. 18, fig. 68, eyes; Pl. 19, fig. 114, lateral view. *Suppl.*]

**4. Synemosyna picata.**

Pl. 9, fig. 21.

*Description.* Black; legs varied with rufous and black, second pair black beneath, fourth black except the knee which is pale beneath; palpi pale, basal joint piceous; feet, 4. 3. 2. 1.

*Observations.* This is evidently distinct from the other species, particularly by its form. I once enclosed a male and a female of this species in a glass tube. They very soon formed separate habitations of silk; but on the third or fourth day the male was dead near the tent of the female, and she had made a lenticular white cocoon, containing four eggs as large as those of large Arancides. That female had a white streak on each side of the abdomen.

*Habitat.* North Carolina, Alabama.

[Taken in June. *Supplement.*]

---

[Continued from Vol. v, p. 370.]

Genus THOMISUS. Walck.

Characters. *Cheliceres small, cuneiform, fang small: maxillæ pointed at tip, more or less inclined over the lip; lip pointed at tip, wider in the middle than at base, as long as, or longer than, half the length of the maxillæ; eyes eight, equal or sub-*

*equal, commonly in two rows of four each, the posterior one longest, bent from the base towards the anterior one; feet, commonly the first and second pair longest, or the second alone longest.*

*Habits.* Araneides wandering after prey, making no web, [444] but casting irregular threads; cocoon flattened, usually placed under leaves, watched by the mother till the young are hatched.

*Remarks.* Well was it remarked by Walckenaer, that a subgenus so easily recognized as Thomisus is nevertheless excessively difficult to characterize. There is not one feature, save the small size of the cheliceres, a secondary character, which is not liable to vary in the different species, and yet, the subdivision is a natural one. Nay, the subgenus Philodromas which seems to correspond to my first tribe, the Depressæ, does not appear to be sufficiently characterized to be separated from this, at least if my *Thomisus vulgaris* can be referred to it.

SECTION I. HETEROPODÆ. *Four posterior legs shortest.*

Tribe 1. DEPRESSÆ. *Legs very long, equal in thickness, body flattened.*

**1. Thomisus vulgaris.**

Pl. 10, fig. 1.

*Description.* Pale gray, abdomen with four impressed dots, body flat; legs with indistinct darker rings.

*Observations.* This spider, commonly seen on fencing or prostrate timber, like those of the same genus, moves sideways and backwards, but it is much more active than *T. celer.* When pursued by an enemy, like Attus and Epeira, it leaps and hangs by a thread, which supports it in the air.

*Habitat.* United States.

[Pl. 18, fig. 77, eyes. Legs arranged 2. 1. 3. 4. This spider withdraws into the chinks of fences, falls attached to a thread, and always moves off sideways. *Supplement.*]

[♀, length 6.8 mm.; cephalothorax 2.6 mm.; legs 11, 12.6, 10.2, 10.
♂ " 6.4 mm. " 2.8 mm.; " 15.5, 17.6, 14.2, 14.2.
Palpus of ♂. Pl. 20, fig. 10.
Salem, Mass. April 6, on fences; May 20, ♂ and ♀.
Providence, R. I; Albany, N. Y.
(Ohio, ♀. Wm. Holden.) J. H. E.]

Tribe II. CANCROÏDES. *Legs very long, four anterior ones largest, abdomen oval.*

### 2. Thomisus aleatorius.

Pl. 10, fig. 2. .

*Description.* Cephalothorax greenish yellow, region of the [445] eyes reddish with whitish lines between and before the eyes, trophi piceous; abdomen yellow with six impressed dots, yellow underneath; feet, two first pair piceous, third and fourth greenish yellow. A small species.

*Observations.* This little spider is not rare, usually found on plants.

*Habitat.* Alabama, September.

[Pl. 18, fig. 39, eyes. Legs arranged 1. 2. $\overline{3.\ 4}$. *Supplement.*]

Tribe III. PYRIFORMES. *Legs moderately long, abdomen pyriform.*

### 3. Thomisus ferox.

Pl. 10, fig. 3.

*Description.* Brownish yellow; cephalothorax with a dusky band each side, abdomen with four or six angular brownish spots; two anterior pair of feet hairy.

*Observations.* This common species is found on plants with the same habits as *T. celer.* I have found in Alabama, in April, a male and a female on a bush; the male was grasping her with his long legs. His abdomen was not truncated, and its marking was somewhat different from that of the female. This spider is apt to vary in color and marking.

*Habitat.* United States.

[Pl. 18, fig. 83, eyes. Legs arranged 1. 2. 3. 4. Taken in Massachusetts in October and March. *Supplement.*]

[Marietta, Ohio, ♂, ♀. Wm. Holden. J. H. E.]

### 4. Thomisus fartus.

Pl. 10, fig. 4.

*Description.* Pale yellow; cephalothorax with an orange fascia in the region of the eyes; abdomen with a marginal red band not reaching the apex, and five or seven impressed dots, the band sometimes obsolete.

*Observations.* This elegant species, first found on the *Actæa spicata* in the Cambridge botanic garden, and often seen since in various places, is always found on plants. It watches its cocoon, which is attached usually under a leaf, [446] like that of a Coccinella, and remains near till the eggs are hatched. It varies in marking, and I have one specimen with red spots on the back of the abdomen. It is nevertheless distinct from *T. celer*, and may be *T. citreus*, Règne An. iv. 256.

*Habitat.* Massachusetts, Alabama.

[Pl. 18, fig. 69, eyes. Legs arranged, 1. 2. 4. 3. *Supplement.*]

[Length of ♀ 10.3 mm.; cephalothorax 3.6; legs 13, 13, 6.6, 7.4.

Milk white or light yellow, sometimes with a crimson stripe each side of the abdomen, and a crimson spot between the eyes.

Salem and Beverly, Mass., throughout the summer; young in May and August. J. H. E.]

### 5. Thomisus celer.

Pl. 10, fig. 5.

*Description.* Pale yellow, with a slight tinge of grass

green, particularly on the legs. Two curved rows of impressed dots on the abdomen; lateral eyes not borne on tubercles.

*Observations.* This spider is found usually on blossoms, where it remains, patiently waiting for Diptera, other small insects, and even butterflies, which its secures with amazing muscular power. It moves backwards and sideways more commonly than forwards. Sometimes seen larger, though never attaining great dimensions.

*Habitat.* Found in South Carolina, North Carolina, Massachusetts, Alabama, Ohio, etc.

[Pl. 18, fig. 78, eyes. Legs arranged $\overline{1.2.4.3.}$ *Supplement.*]

[Marietta, Ohio, ♀. Charlestown, Mass. Wm. Holden. J. H. E.]

Tribe IV. OCULATÆ. *External eyes borne on tubercles, eyes equal.*

### 6. Thomisus piger.

Pl. 10, fig. 6.

*Description.* Yellowish brown; cephalothorax with two brownish bands; abdomen with two curved lines of impressed dots; somewhat paler underneath.

*Observations.* This species is probably the largest, and is very distinct from any other, particularly by its habits. It dwells under stones, where it watches for its prey, and has not been found on plants.

*Habitat.* North Carolina. [447.]

[Pl. 18, fig. 40, eyes. Legs arranged $\overline{1.2\ 4.3.}$ Taken in March. *Supplement.*]

### 7. Thomisus asperatus.

Pl. 10, fig. 7.

*Description.* Pale, covered with short bristles; cephalothorax, outer eyes of second row tuberculated, with two blackish bands, and a few longer bristles about the eyes; abdomen pale brown above, pale glabrous underneath; feet, first and second pair with brown rings, third and fourth with fewer bristles.

*Observations.* This spider, which is found on plants, seems quite distinct from any other, though the species of this subgenus are very variable.

*Habitat.* Alabama. September.

[Pl. 18, fig. 41, eyes. Legs arranged $\overline{1. 2. 3. 4.}$ *Supplement.*]

### 8. Thomisus parvulus.

Pl. 10, fig. 8.

*Description.* Rufous; abdomen yellowish with a transverse band black, near the apex; third and fourth pair of legs greenish or yellowish; first and second pair longer and slender in the male.

*Observations.* The external eyes appear to be larger, on account of their being placed on elevations. This species is common, and frequently found on the blossoms of umbelliferous plants.

*Habitat.* The Southern States.

[Pl. 18, fig. 42, eyes. The ♂ resemblesthe ♀, and is of the same size; the legs are green in the ♀ and pale in the ♂; legs arranged $\overline{2. 1. 4. 3.}$ Taken in Carolina May 25. *Supplement.*]

Tribe V. TUBERCULATÆ. *A tubercle on the abdomen, external eyes larger, tuberculated.*

### 9. Thomisus caudatus.

Pl. 10, fig. 9.

*Description.* Dusky; abdomen with a tubercle or tail behind above the apex, about six small black dots on the disk; pale beneath, with a longitudinal band, and sides blackish; feet, 1. 2. $\overline{4. 3.}$ [448.]

*Observations.* This species is not rare. It is sometimes found wandering in mid-winter. The eyes are unequal in size, the two lower external ones are largest, and the four external ones are borne on tubercles.

*Habitat.* Alabama.

[Pl. 18, fig. 60, eyes; fig. 100, trophi. Taken on January 25. *Supplement.*]

[♀, length 5.8 mm.; cephalothorax 2.4 mm.; legs 9.6, 9.6, 5.4, 6.

Beverly, Mass.; June 23, with cocoon folded in leaf. Peabody, Mass.; Sept. 28, young. J. H. E.]

SECTION II. ÆQUIPEDES. *Four posterior legs not invariably the shortest.*

Tribe VI. FILIPEDES. *Feet slender, long, second pair longest, then the fourth.*

### 10. Thomisus? Duttoni.

Pl. 10, fig. 10.

*Description.* Pale gray; cephalothorax with a longitudinal rufous band; abdomen long and slender, with a like narrow band, and two minute black dots near the apex; legs yellowish, 2. 4. 1. 3.

*Observations.* This singular spider was communicated to me by Mr. Thomas R. Dutton, who collected it in Georgia in 1838. The alcohol in which it was preserved may have changed its colors.

*Habitat.* Georgia.

[Pl. 18, fig. 79, eyes. *Supplement.*]

[♀, length 11 mm.; cephalothorax 3 mm.; legs 12, 13.6, 8.7, 12.4.

♂    "    6 mm.;      "    2.4 mm.; "    10.8, 11.6, 8.3, 11.

Palpus of ♂. Pl. 20, fig. 11.

Beverly, July 14; Malden, Moody; Mass., Sanborn. Dakota Terr., E. Coues. J. H. E.]

Tribe VII. not determined; *eyes in four rows.*

## 11. Thomisus? dubius.

Pl. 10, fig. 11.

*Description.* Pale; cephalothorax with two slender longitudinal blackish lines edged with greenish; abdomen with a similar green edged line, which bifurcates towards the base, and has one small black dot on each bifurcation; feet, first pair wanting, second very long, fourth next, third shortest.

*Observations.* This singular spider was unfortunately mutilated when discovered, and the drawing was left unfinished, as I hoped other specimens would occur; none, however, were ever found. There is an affinity in some points between this and *T. Duttoni.*

*Habitat.* North Carolina. [**449.**]

[Taken in April. *Supplement.*]

## 12. Thomisus? tenuis.

Pl. 10, fig. 12.

*Description.* Testaceous, downy; cephalothorax with a longitudinal white band, and a tuft of hair between the eyes; abdomen with two interrupted longitudinal whitish fillets, four long nipples; feet bristly, 2. $\overline{1.~4.}$ 3.

*Observations.* This is undoubtedly congeneric with my *Thomisus dubius.* But knowing nothing of their webs nor of habits, I still refrain from making any generic distinction. These will probably form the type of a new sub-genus. This one was found enclosed in the clay nest of a Spex.

*Habitat.* Alabama.

[Pl. 18, fig. 84, eyes; fig. 101, trophi. The two anterior eyes are placed on tubercles on the very margin.; body beneath yellowish, downy. Taken June 8. *Supplement.*]

## Genus CLUBIONA. Latr.

Characters. *Chelicercs long, fang moderately long; maxillæ parallel, wider above the insertion of the palpi, lip widest in the middle; eyes eight, equal, in two rows, the lower one nearly straight; feet, the fourth or the first pair longest; body usually of a pale or livid color.*

*Habits.* Araneides sedentary, watching their prey, and inclosing themselves in silk tubes ; cocoon orbicular.

*Remarks.* This sub-genus, like Thomisus, cannot be characterized with any precision, owing to the variations in the form of the trophi, the position of the eyes, etc. It is nevertheless a natural subdivision of Aranea. The species composing it have nocturnal habits; little therefore, is known of their history. They dwell under leaves, under bark or stones, where they may be found in silk tubes, from which they seldom issue during the day.

Tribe I. DRYADES. *Eyes, posterior row bent toward the base; fourth pair of legs longest, then the second.*

### 1. Clubiona pallens.

Pl. 10, fig. 13.

*Description.* Livid white ; abdomen varied with plumbeous [450] spots above, and four small dots near the apex underneath ; feet, 4. $\overline{2}$. $\overline{1}$. 3. Both sexes alike.

*Observations.* This spider is found in silk tubes, concealed under the bark of decaying trees, where it spends the winter. There is a spotless variety which may prove a distinct species.

*Habitat.* North Carolina, Alabama ; common.

[Pl. 18, fig. 7, eyes. The second pair of legs is always sensibly longer than the first. Taken December 15. *Supplement.*]

[♀, length 6.8 mm.; cephalothorax 3.2 mm. ; legs 6.7, 7, 6, 9.
♂ " 6 mm.; " 3 mm.; " 7, 7.4, 6, 8.6.
Palpus of ♂. Pl. 20, fig. 13.

The original drawing has the cephalothorax light yellowish-brown, darkest towards the eyes. The rest of the body is yellowish-white, with gray markings as described. Young specimens have the whole body yellow-white with gray markings.

Salem, Mass. March, under stones in bags; torpid.

Dedham, Mass. January 9, under leaves.

Providence, R. I. J. H. E.]

## 2. Clubiona obesa.

Pl. 10, fig. 14.

*Description.* Testaceous or brownish, abdomen with a longitudinal more or less distinct brown band above.

*Observations.* This spider, usually found concealed in silk tubes, was sometimes seen in the blossoms of the Magnolia seeking for prey. It is perfectly distinct from *Clubiona inclusa.*

*Habitat.* Massachusetts, North Carolina, Alabama.

[Pl. 18, fig. 16, eyes. Legs arranged 4. 2. 1. 3. Found hanging from trees by a thread. Taken at the end of June. *Supplement.*]

[♀, length 12.5 mm.; cephalothorax 4.3 mm.; legs 10.8, 10, 9.7, 12.
♀ " 8.4 mm.; " 3.6 mm.: " 10, 9, 8.4, 11.2.
♂ " 9 mm.; " 3.7 mm.; " 13.9, 12.2, 10.2, 13.4.
Palpus of ♂. Pl. 20, fig. 12.

The figure, pl. 23, fig. 14, represents a ♀ with unusually large abdomen. Salem. March, young females under stones. April 6, under stones. May 24, under stones, ♂ and ♀ enclosed together in a thin silk bag an inch in diameter. June 16, ♂ and immature ♀ in a bag together; young ♀ in bags with cast skins. July 7, in thin bag with cocoon of thirty-three eggs. July 12, ♀ confined in a bottle laid eggs, which hatched Aug. 5, and left the cocoon Aug. 26. Sept. and Oct., under stones and logs in woods, Ann Arbor, Mich. J. H. E.]

Tribe II. HAMADRYADES. *Eyes, posterior row bent from the base; lip emarginate; first pair of legs longest.*

## 3. Clubiona piscatoria.

Pl. 10, fig. 15.

*Description.* Dingy rufous; eyes sub-equal, two middle

ones larger; abdomen pale, piceous, with a sub-obselete spot near the base, four nipples, two external ones bi-articulate. Feet long, 1. 4. 2. 3.

*Observations.* The difference between the eyes of this and those of my *Clubiona obesa* prevents my referring them to the same species. The pulmonary orifices are white, under a gloss. This spider made an even web like Agelena. Wandering at night.

*Habitat.* Alabama.

[Pl. 18, fig. 20, eyes. The two external nipples of the abdomen are the longest; the body is of the same color beneath as above. Taken in April. *Supplement.*]

### 4. Clubiona tranquilla.

Pl. 10, fig. 16.

*Description.* Deep rufous or piceous; abdomen grayish black, with four impressed dots. [451.]

*Observations.* It is difficult to learn much of the habits of this spider, which moves chiefly at night. A male and a female were found in Alabama in July, in the folds of an old piece of paper, near a silk tube of extreme whiteness, which was probably destined to receive the eggs. Always found in a tube except at the approach of winter, when it is sometimes found wandering.

*Habitat.* Common in the United States.

[Pl. 18, fig. 85, eyes; fig. 102, trophi. Legs arranged 1. 2. 4. 3. Taken in Alabama and California. *Supplement.*]

Tribe III. NYMPHÆ. *External eyes approximated, lip emarginate, first pair of legs longest.*

### 5. Clubiona inclusa.

Pl. 10, fig. 18.

*Description.* Livid white, or pale yellow; cheliceres, last joints of all the feet and of the palpi tipped with black; a longitudinal dusky line beginning at base of the abdomen.

*Observations.* This spider was always found in tubes of white silk, the female watching her cocoon, which is covered with a very thin coat of silk; the eggs are loose and not glued together. It probably moves out only at night, as its pale color indicates. The young are deeper in color even than the mother.

*Habitat.* South Carolina, North Carolina, etc.

[Pl. 18, fig. 86, eyes. Legs arranged 1. 4. 2. 3. Taken in June. *Supplement.*]

[Charlestown, Mass., ♀. Hyde Park, Mass., ♀. Wm. Holden. J. H. E.]

Tribe IV. Furlæ. *External eyes not touching, lip truncated at tip, fourth pair of legs longest.*

### 6. Clubiona fallens.

Pl. 10, fig. 17.

*Description.* Yellowish or rufous; cephalothorax with blackish lines; abdomen pale, with two rows of sub-obsolete dots, and two abbreviated rows of smaller ones obscure, same color underneath; feet hairy, particularly the third and fourth pair, in the female, 4. 1. 2. 3. and in the male $\overline{1.\ 4}$. 2. 3. The sexes marked alike.

*Observations.* Were it not that the eyes are differently [452] placed, this might be taken for *C. celer.* Males and females were found in silk tubes constructed on leaves. A male was found in November, with imperfect blackish rings on the legs, one more distinct at the base of the antepenultimate joint. Is it a variety or a distinct species?

*Habitat.* Alabama, October, November.

[Pl. 18, fig. 17, eyes. *Supplement.*]

### 7. Clubiona gracilis.

Pl. 10, fig. 19.

*Description.* Yellowish; cephalothorax with two longitudinal bluish bands; abdomen with two longitudinal bands of numerous small red dots, the bands uniting towards the apex; feet, hairy, 4. 1. 2. 3. The sexes alike.

*Observations.* This very active spider is often seen in midwinter on a mild day apparently migrating in great numbers, being supported in the air by a long thread, and borne by the breeze. Once, many were seen in December, thus approaching a large tree, under the bark of which they probably intended to hibernate. A variety, perhaps a distinct species, was found, destitute of dots or bands; it was concealed in a silk tube on a leaf.

*Habitat.* North Carolina, Alabama.

[Pl. 18, fig. 8, eyes. Taken in June and July. *Supplement.*]

[♀, length 8.8 mm.; cephalothorax 3 mm.; legs 9.4, 7.6, 6.2, 9.8. Mandibles and front of head blackish-brown.

Saugus, Mass.; June 12. Boston, Mass.; October, flying for fences. Providence, R. I.; October. New Haven, Conn. Freehold, N. J.; S. Lockwood. J. H. E.]


**8. Clubiona celer.**

Pl. 10, fig. 20.

*Description.* Pale; cephalothorax with angular markings near the edge, and some lines; abdomen pubescent, with indistinct dots; feet, hairy. Male resembling the female in every point of markings.

*Observations.* This little spider was found in December, suspending itself from a thread, and moving with great activity. A male somewhat larger than the scale was found in Alabama, April, wandering about at night.

*Habitat.* North Carolina, Alabama.

[Pl. 18, fig. 18, eyes. Legs arranged 1. 4. 2. 3. Taken in June. *Supplement.*]

Tribe V. CAVERNOSÆ. *Eyes in two sub-parallel rows, lip pointed at tip, feet variable.* [453.]

### 9. Clubiona? agrestis.

Pl. 10, fig. 21.

*Description.* Livid green; abdomen purplish brown, with four impressed dots; feet, 1. 4. 2. 3. Male of a piceous color.

*Observations.* The male and the female were found under a stone. The female has two curved rigid shining elevations under the vulva, forming an arch open towards the base. It is with some hesitation that I place this species in this subdivision.

*Habitat.* Alabama. March.

[Pl. 19, fig. 43, eyes. *Supplement.*]

### 10. Clubiona immatura.

Pl. 10, fig. 22.

*Description.* Yellowish rufous, middle lower eyes black; abdomen immaculate pale green; legs with very short hairs, 4. 1. 2. 3.

*Observation.* This was found in a cellar.

*Habitat.* Alabama. October.

[Pl. 18, fig. 87, eyes. *Supplement.*]

Tribe VI. Not determined.

### 11. Clubiona? sublurida.

Pl. 11, fig. 1.

*Description.* Pale yellowish; cheliceres very large; abdomen with two obscure sub-obselete lines, same color beneath with a few minute brown spots; feet, long, slender, 1. 4. 2. 3.

*Observations.* This spider was found upon a bush, without any web. It displayed great activity and vigor. When at rest it had its legs spread out.

*Habitat.* Alabama. July.

### 12. Clubiona? saltabunda.

Pl. 10, fig. 23.

*Description.* Pale; cephalothorax with a few obscure marks near the edge, palpi with bristles; abdomen with two [454] rows of blackish dots, and a few minute ones towards the sides; pectus with a scalloped black line on each side; venter with four or five small spots, and many small dots black; feet, slender, 1. very long, 4. 2. 3.

*Observations.* This spider is found in the fields, wandering, and running with great activity. It leaps like Attus, and like it too, it leaves a thread behind to secure its flight. A female found in November, made a tube or tent as a residence in the vial in which it was enclosed. The male resembles the female. Probably congeneric with *Clubiona? subhurida.*

*Habitat.* Alabama. May, November.

[Pl. 18, fig. 19, eyes. Prof. Hentz was in doubt whether this was a Clubiona or a Tegenaria. *Supplement.*]

[♂, length 3 mm.; cephalothorax 1.5 mm.; legs 9.8, 5.4, 4.6, 6.3. Palpus of ♂. Pl. 20, fig. 15.
West Roxbury; June 2, ♂. Peabody, Mass.; June 29, ♂. Waverly, Mass.; Oct. 3, young ♀. J. H. E.]

### 13. Clubiona? albens.

Pl. 10, fig. 24.

*Description.* Pale bristly; abdomen deeper in color, venter with its base and three spots, pale green, pectus with a line each side, pale green, nipples, four long ones and two short. Feet, very long and slender, 1. 4. 2. 3. First much the longest.

*Observations.* This is undoubtedly related to my *C. salta-bunda*, and with it will probably constitute a new subgenus at some future time.

*Habitat.* Alabama.

[Pl. 18, fig. 32, eyes. Found travelling from one bush to another by means of a thread. Taken April 15. *Supplement.*]

### Genus HERPYLLUS. Mihi.

Silliman's Jour., Vol. XXI, p. 102.

*Characters.* *Cheliceres moderately large, without teeth; maxillæ parallel, wider above the insertion of the palpi, cut obliquely above; lip about half the length of the maxillæ, narrower towards the point; eyes, eight, sub-equal in two parallel rows of four each, both commonly bent towards the base; feet, the fourth pair longest, then the first, then the second, the third being the shortest.*

*Habits.* Aranéides wandering after prey, making no web, but running about with great swiftness, and hiding under stones, in crevices, etc. Cocoon unknown. [455.]

*Remarks.* This sub-genus, very closely related to Clubiona, is very well characterized notwithstanding the similarity. The character derived from the respective length of the legs is very constant, showing the property of being fast runners in all the species of this division.

In habits they differ wholly from Clubiona, being in fact wandering Araneides. The swiftness with which they run is truly surprising. They are not exclusively nocturnal, being often seen to run in the brightest sunshine.

Tribe I. BREVIPEDES. *Legs strong, rather short, maxillæ long, cut obliquely above.*

#### 1. Herpyllus ecclesiasticus.

Pl. 11, fig. 2.

*Description.* Black; cephalothorax with a whitish longitudinal band; abdomen with an abbreviated band, and a spot white.

*Observations.* This spider is not rare, found between boards and in crevices in dark places; running very fast, chiefly at night; I never could find its cocoon or its permanent dwelling-place.

*Habitat.* The United States.

[Legs arranged 4. 1. 2. 3. This, and others placed in this genus, may belong to the genus Diplotoxops of Mr. Rafinesque, but as he makes the first pair of legs longest, and his generic description is incorrect in many respects — for instance, in deriving a character from the palpi, which is as a rule nothing but a sexual distinction — his name has not been adopted. *Supplement.*]

[Ohio, ♂, ♀; Mayport, Fla., ♀. The stripe on the abdomen pale or wanting. Wm. Holden. J. H. E.]

## 2. Herpyllus ater.

Pl. 11, fig. 3.

*Description.* Deep glossy black, immaculate, feet rather short.

*Observations.* This species, readily distinguished from *H. descriptus,* which has long slender legs, is found running with great rapidity on paths and frequented places. It is diurnal, and when pursued it seeks shelter under stones or leaves. It has the same habits with *H. bicolor,* to which it is related.

*Habitat.* Pennsylvania, New England. [456.]

[♀, length 7.2 mm.; cephalothorax 3.2 mm.; legs 7.2, 6, 6, 8.4.
♂, " 6.2 mm.; " 2.5 mm.; " 5, 4.6, 4, 6.
Palpus of ♂. Pl. 20, fig. 16, 16a.
Salem, Mass. April 19, in a bag under stones. July 30, under stones. Sept. 29, in a bag under stones. West Roxbury ; June 1, ♀ in a thin bag with a flat, pink cocoon. White Mountains, Sanborn. Marietta, Ohio, ♀. Wm. Holden. J. H. E.]

## 3. Herpyllus bicolor.

Pl. 11, fig. 4.

*Description.* Rufous; abdomen bluish black, with about six impressed dots. Male same colors.

*Observations.* This common species is found usually on the ground or under stones, leaves, etc., running with great rapidity. The female almost invariably kills the male and eats him, after the calls of nature are satisfied. It is related to *H. ater.*

*Habitat.* North Carolina, Alabama, and probably the United States.

[Ohio, ♂, ♀. Wm. Holden. J. H. E.]

#### 4. Herpyllus bilineatus.

Pl. 11, fig. 5.

*Description.* Whitish; cephalothorax above, and abdomen above and beneath with two longitudinal black bands somewhat curved; feet yellowish. Male with the same marks.

*Observations.* This spider is remarkably active, usually found on trees. No species of this sub-genus is more distinct, and invariably spotted in the same manner. It is not rare.

*Habitat.* North Carolina, Alabama.

[Of the six nipples, four were placed around the anus, and two formed a fork on both sides of the anus. Taken in May. *Supplement.*]

#### 5. Herpyllus ornatus.

Pl. 11, fig. 6.

*Description.* Golden rufous; abdomen with abbreviated and interrupted transverse bands black; feet, yellowish, varied with black, thighs of two anterior pairs of legs black.

*Observations.* Wandering on paths, and very active, in woods or unfrequented places. The young usually has its abdomen black, with transverse whitish bands which are formed by hairs.

*Habitat.* North Carolina.

[It is difficult to catch. Taken in July and August. *Supplement.*]

#### 6. Herpyllus descriptus.

Pl. 11, fig. 7.

*Description.* Black; abdomen with an abbreviated, longitudinal [457] band, golden rufous; two anterior pairs of legs with last three joints brownish or paler.

*Observations.* There are probably several species very similar to this. A large specimen was found in Alabama, with no yellow spot on its abdomen, its legs with a few stout bristles, its two posterior thighs had two whitish bands above, and the base of its abdomen had a whitish spot, these bands and spot formed by short hairs. Another specimen occurred with the whole disc of the abdomen red, the abdomen had a peduncle one-third the length of the cephalothorax, the cheliceres were more prominent. Are these two distinct species? This is closely related to *H. ornatus.*

*Habitat.* North Carolina, Alabama.

[♀, length 8.7 mm.; cephalothorax 3.4 mm.; legs 7.2, 7.2, 7.2, 10.2.
♂,  " 7.4 mm.;  "  3.4 mm.;  " 7.8, 7.2, 7.2, 10.7.
Palpus of ♂. Pl. 20, fig. 18.
Salem, Mass.; July 22.  Aug. 5, ♂ under stone ; Sept. 6.  Ohio, ♀.
Wm. Holden. J. H. E.]

### 7. Herpyllus crocatus.

Pl. 11, fig. 8.

*Description.* Piceous black ; abdomen darker, with a saffron-colored band widening towards the apex, blackish beneath ; feet, fourth pair hairy.

*Observations.* This species inhabits houses, hiding in cracks, under boards, etc. It does not vary in marking, and is very well characterized.

*Habitat.* Alabama.  November.

[Zanesville, Ohio, ♀.  Charlestown, Mass., ♂.  Wm. Holden.  J. H. E.]

### 8. Herpyllus longipalpus.

Pl. 11, fig. 9.

*Description.* Black ; palpi nearly as long as the cephalothorax ; abdomen with sub-obsolete transverse white bands ; feet spotted with white ; immaculate black underneath.

*Observations.* The spots and bands in this, as well as in most of the other species of this sub-genus, are produced by scales or hairs which are quite deciduous, and hence there are many varieties of markings. This spider moves with the rapidity of lightning.

*Habitat.* Alabama. September. [**458.**]

[Legs arranged 4. 1̄. 2̄. 3̄. *Supplement.*]

### 9. Herpyllus marmoratus.

Pl. 11, fig. 10.

*Description.* Black, varied with whitish markings formed by deciduous scales; feet, 4. 1. 2. 3. Fourth pair stoutest.

*Observations.* This can scarcely be the male of my *H. longipalpus.*

*Habitat.* Alabama.

[First, second, and third pairs of legs pale, thighs black, with the tip white, third and fourth with bands of white scales. Taken in July. *Supplement.*]

### 10. Herpyllus variegatus.

Pl. 11, fig. 12.

*Description.* Cephalothorax rufous; abdomen blackish, with three whitish bands, the middle one as an inverted ⊥; feet, varied with piceous and rufous.

*Observations.* This spider, drawn from a specimen collected by Prof. Peck of Massachusetts, was immersed in spirits ten or twelve years at least, before it was painted. The colors may not be correctly represented in consequence of that. A specimen was found in North Carolina, and also one in Kentucky, in a silk tube, which had only two bands on the abdomen, and the external eyes of which were placed nearer together. These will probably be found to belong to another species.

*Habitat.* Massachusetts?

[Specimens from North Carolina and Kentucky had the ex-

ternal eyes of the posterior line more advanced toward those of
the anterior line. These specimens were also smaller. *Supplement.*]

[♂, length, 6.5 mm.; cephalothorax 2.8 mm.; legs 5.4, 5, 5, 6.6.
Palpus of ♂. Pl. 20, fig. 17, 17a.
Maklen, Mass., II. L. Moody. J. II. E.]

### 11. Herpyllus cruciger.

Pl. 11, fig. 11.

*Description.* Gray; abdomen with spots and dots black.
*Observations.* This spider is really black, but covered with
gray hairs or scales which can be easily rubbed off, and which
are arranged on the abdomen somewhat in the form of a cross.
It moves with great celerity, and hides under stones, etc.
*Habitat.* North Carolina.
[Taken in July. *Supplement.*]

### 12. Herpyllus vespa.

Pl. 11, fig. 13.

*Description.* Piceous; cephalothorax with the middle [459]
lower eyes black; abdomen usually deeper in color, with four
impressed dots, separated from the cephalothorax by a whitish
peduncle, underneath with a pale spot over each pulmonary
orifice.
*Observations.* This spider, like other congeneric species,
runs very fast and conceals itself under stones. It is common.
It may be that *Agelena plumbea* will be referred to this.
*Habitat.* Alabama.
[Taken in March. *Supplement.*]

### 13. Herpyllus ? ramulosus.

Pl. 11, fig. 14.

*Description.* Obscure brown; abdomen with two diverging
bands and several spots pale brown, spotless pale beneath.

*Observations.* This may be referred to Clubiona, as it bears some affinity to *C. celer* and others.

*Habitat.* Alabama. May.

### 14. Herpyllus? pygmæus.

Pl. 11, fig. 16.

*Description.* Piceous; feet and palpi paler; feet, 4. $\overline{2. 1. 3}$. A very small species.

*Observations.* This species is referred to this division with but little hesitation. It is probably not common. It was found wandering.

*Habitat.* Alabama. August.

Tribe II. LONGIPEDES. *Legs slender, long, maxillæ short, truncated.*

### 15. Herpyllus? auratus.

Pl. 11, fig. 15.

*Description.* Bright rufous; abdomen brilliant gold color above and beneath, with four abbreviated white lines above, and four on the sides towards the base, with a tinge of silvery green around the vulva in the female; feet, filiform, long and slender, dusky towards the extremity, particularly the fourth pair. [460.]

*Observations.* This beautiful slender species moves like a mouse, and with such rapidity as to make it quite an arduous undertaking to capture it. The male and female have been repeatedly found with the same colors and marking. One specimen, soon after being inclosed in a glass tube, made a beautifully wrought tent like that of Attus, open at both ends. It would seem that this spider has a fixed place of abode, from which it issues for hunting excursions, for a female was observed by some children, several times on the same plant, repeatedly escaping to the ground when pursued, until it was at last taken in the very same spot. A female in a state of gravidity was found September 30th, agreeing in every description except in having obscure bands in the form of an ∧; about four distinct ones, near the apex.

*Habitat.* Alabama. August, October.

### 16. Herpyllus zonarius.

Pl. 11, fig. 17.

*Description.* Brown; abdomen piceous, with two transverse white bands interrupted in the centre, unspotted beneath; feet varied with brown and yellowish. A small species.

*Observations.* This little spider is probably not a variety of *H. auratus.* Its feet are not so filiform. It runs with great celerity.

*Habitat.* Alabama. September.

[Ohio, ♂, ♀. Wm. Holden. J. H. E.]

### 17. Herpyllus trilineatus.

Pl. 11, fig. 18.

*Description.* Rufous; abdomen with three transverse golden yellow lines or bands produced by hairs, rufous unspotted beneath; feet, slender and long, paler towards the extremity, penultimate joint blackish, particularly of the first and second pair. Both sexes alike.

*Observation.* This spider was found wandering.

*Habitat.* Alabama. April, May. [461.]

[Athens, Ohio, ♀. Wm. Holden. J. H. E.]

### 18. Herpyllus parcus.

Pl. 11, fig. 19.

*Description.* Rufous; abdomen with some transverse subobsolete obscure bands near the apex, where the abdomen is covered with hair which turns pale green in a certain light, pale underneath; first two pair of legs with two rows of knobs on which long hairs are inserted. A small species.

*Observations.* This spider is usually found under logs in the woods. It is strongly characterized, and cannot be taken for any other. The hairs or bristles on the knobs of the legs are laid close on the leg, and are not visible to the naked eye on that account; they are probably susceptible of voluntary motion, for defence.

*Habitat.* Alabama. July, September.

### 19. Herpyllus alarius.

Pl. 11, fig. 20.

*Description.* Cephalothorax pale rufous, with a scolloped margin darker; abdomen obscure piceous with four or five transverse bent lines yellowish; feet pale, first pair with the top of the thighs and the two next joints blackish hairy, second pair with a blackish ring on the antepenultimate joint. A small species.

*Observations.* This species was found under a board.

*Habitat.* Alabama.

[♀, length, 4 mm.; cephalothorax, 1.5 mm.; legs 4.4, 4, 3.8. 5.4.
♂ " 2 mm.; . " 1 mm.; legs 3.4, 2.6, 2.4, 4.
Pl. 21, fig. 14, palpus of ♂.
Salem, Mass., March; young, under stones. Gloucester, Mass., Aug. 17; Brookline, Jan. 24; under leaves. Providence, R. I., May 28, ♂. Zanesville, Ohio, ♀, Wm. Holden. J. H. E.]

Tribe III. Doubtful.

### 20. Herpyllus? dubius.

*Description.* Black; abdomen with two white spots; feet rufous, thighs black.

*Observations.* This species, unfortunately not completely painted, was found running on walls.

*Habitat.* South Carolina. [462.]

### Genus TEGENARIA. Latr. Walck.

Characters. *Chelicerae moderately long; maxillae parallel, rounded, very slightly inclined towards the lip; lip short, rounded at tip ; eyes eight, equal, in two rows, anterior one composed of four eyes in a straight line, posterior one longer, curved towards the base ; feet, fourth pair longest, then the first, the other two nearly equal.*

*Habits.* Araneides sedentary, making in obscure corners an horizontal web, at the upper part of which is a tubular habitation where the spider remains motionless till some insect be entangled in the threads.

*Remarks.* The distinction first proposed by Latreille between this and the Agelena of Walckenaer should be preserved. The habits of the spider differ considerably, and the position of the eyes is so different as to point out the necessity of a separation. I would have preserved the name Aranea to this division ; but confused ideas would arise from attributing to a sub-genus the name, which, though legitimate, belongs more properly to the whole family of spiders.

These make webs of slender texture in dark places without the addition of the strong cross threads which Agelena adds to the horizontal texture. It is only at night that they can be seen at work in the construction of their webs.

### 1. Tegenaria medicinalis.

Journal of the Acad. of Nat. Sci. Philad., II, p. 53, pl. v, fig. 1.

Pl. 11, fig. 21.

*Description.* Pale brown ; turning to bluish black ; cephalothorax with a blackish band on each side ; abdomen varied with black, or plumbeous and brown ; feet varied with blackish.

*Observations.* This species, which was described by the author in the Journal quoted above, is found in every cellar or dark place in the country. For some time the use of its web [463] as a narcotic in cases of fever, was recommended by many physicians in this country ; but now it is probably seldom used. The author being absent from Philadelphia when the second volume of the Journal was published, a strange mistake was committed. The publishers caused a delineation of my

*Lycosa lenta* to be printed instead of the original drawing of this species, which was probably lost ; and as soon as they were informed of the error, they caused an imperfect delineation of this species to be substituted, which may be seen on Plate v, along with some representations of crystals of Zircon, published by Dr. G. Troost. The palpi of the male of this spider are very complicated, as may be seen by the drawing. The colors vary much. [The article from the Philadelphia Journal will be found further on.]

On the 28th of February, I observed a male specimen of this species in a dark corner, apparently devouring another spider. On moving them with a straw I discovered that the other was a female of the same species, and not dead, but with its legs closely folded on its body, and perfectly motionless. One of the palpi of the male was buried in the vulva of the female, and could not be extricated by the efforts which he made to avoid my intruding straw. I threw them on the ground and had ceased to watch them, when suddenly I saw the female escaping from him, apparently in great terror. In the meanwhile, the male, from whose cheliceres she had escaped, had seized a small bit of stick as a substitute, and ran about with ludicrous haste, seemingly out of his senses for some time. This fact may prove that the female of spiders is not always the tyrant and oppressor of the other sex.

*Habitat.* The United States.

[Pl. 18, fig. 110, palpus of ♂, *a*, upper hook; *b*, lower hook and its membranous meatus above ; *c*, middle sphenoidal piece ; *d*, third hook corresponding to the fourth hook *i*; *e*, bristle which is usually curled between the first and second hook ; *f*, base ; *i*, fourth hook. The upper and lower hooks are parallel, like two fingers ; whilst the third and fourth are opposed to them ; all forming a compound hook which must retain the female organs whilst the little bristle, *e*, stirs and conveys the sperma. That fluid flows from an orifice at the base of the lower hook, between it and the membrane situated on its upper part. The middle piece articulates with all the other parts except the base. The bristle and the third hook articulate together and with the base. The fourth hook seems to be a process of the upper hook. The lower hook articulates with the middle piece, and seems to have no motion of its own. The upper hook articulates with the lower piece. Taken in North Carolina in March, and in Massachusetts in May. *Supplement.*]

[♀, length 11.2 mm.; cephalothorax 5 mm.; legs 14, 12.4, 11.1, 15.
♂ " 9.8 mm.; " 5 mm.; legs 17.5, 15, 14.7, 20.
Pl. 20, fig. 19, palpus of ♂.
The palpi of one of Hentz's specimens are preserved. I have never found this species in houses.
Swampscott, Mass., May 8, under a stone; ♂ and ♀ in copulation.
White Mts.; Schoharie, N. Y.; New Haven, Conn.; Providence, R. I. J. H. E.]

### 2. Tegenaria persica.

Pl. 11, fig. 23.

*Description.* Pale gray; cephalothorax with serrated black lines; abdomen obscure, with about eight pale oblique spots, central line blackish, upper mammulæ very long, obscure beneath, with indistinct markings; feet varied with many [464] blackish rings. Male not differing from the female; feet, 4. 1. or 1. 4. 2. 3. A small species.

*Observations.* This is quite distinct from *T. medicinalis*, by its size, markings, and particularly by the respective length of the legs, the first pair of which is very frequently as long as, or longer than, the fourth. It makes its web on the trunk of trees, with a winding tube turned downward, very much like that of Agelena. I often found it on the peach tree. It never was found larger than the mark on the plate.

*Habitat.* Alabama. September.

### 3. Tegenaria ? flavens.

Pl. 11, fig. 22.

*Description.* Yellowish; cephalothorax rufous; abdomen long and slender; feet slender, 4. 1. 2. 3.

*Observations.* I do not remember where this was found, and it would be well to know what web it makes. It has all the characters of Tegenaria. It must have been some time in whiskey, and the color may have changed.

*Habitat.* Alabama.

### Genus AGELENA. Walck.

*Characters.* *Cheliceres strong; maxillæ slightly inclined, rounded externally; lip conical, as long as, or more than half the length of the maxillæ; eyes eight, equal, two anteriorly, four in a row curved anteriorly, two behind the intermediate ones of the second line; feet, fourth pair longest, then the first, then the second, the third being the shortest, upper mammulæ very long.*

*Habits.* Araneides sedentary, making in the fields, on bushes or stumps, a large horizontal web, with a tubular habitation, the web connected with strong crossed threads extending high above it.

*Remarks.* The name of Walckenaer is preserved for the reasons given in the remarks upon Tegenaria.

No spider is more common or familiar to the eye of every [465] one who rambles in the fields than the first species of this sub-genus. Its habitus is totally different from that of Tegenaria; it is very voracious, attains an immense size, and probably lives many years.

### 1. Agelena nævia? Bosc.

Pl. 12, fig. 1, 1a, young.

*Description.* Rufous hairy, cephalothorax with two longitudinal black bands, abdomen blackish, with two longitudinal rows of whitish dots. Feet very hairy, with joints terminated by a blackish ring.

*Observations.* This species, common in the United States, makes a large horizontal web, spread on bushes or on the grass, with a tubular retreat which terminates in some crevice in the ground, a stump, or any convenient hole to hide itself; strong cross threads are attached to the bushes above the web. It varies very greatly in size, and is remarkably voracious. When very young it makes its web on the ground, on highways; and in the morning, when the earth is covered with dew, myriads can be seen in April and May.

*Habitat.* Common in all parts of the United States.

[♀, length 15 mm.; cephalothorax, 7 mm.; legs 25.2, 22.7, 22.3, 28.8.
♂ " 9.2 mm.; " 4.6 mm.; legs 19.4, 17.5, 17, 20.8.
Pl. 20, fig. 20, palpus of ♂.

Salem, Mass., March 29, young in cocoon after first moult; Apr. 28, dead females under bark with cocoons of young; June 16, young spinning webs in grass. Rowley, Mass., July 14, ♂ on web. Peabody, Mass., Sept. 4, ♂ and ♀; females with cocoons of eggs on leaves of blackberry; Sept. 8 and 22, in copulation in webs. Providence, R. I.,; Portland, Me.; Ann Arbor, Md.; Indianapolis, Ind.

Ft. Cobb, Indian Territory, ♀; very common throughout the west and south; probably two or more species are included under this name. Wm. Holden. J. H. E.]

## 2. Agelena? plumbea.

Pl. 12, fig. 2.

*Description.* Pale rufous; abdomen leaden color, with four impressed dots, the six nipples long; same color beneath, one pale spot each side of the base of the abdomen, over the pulmonary orifices.

*Observations.* This spider was found in North Carolina under a stone, in a silken tube. Another specimen was also found under a stone in Alabama; it was discovered watching a cocoon made of thin but strong white silk, containing about fifty or sixty eggs of a whitish color. As its web was not seen, it may not belong to this division, and may be ultimately referred to Herpyllus.

*Habitat.* North Carolina, Alabama. [466.]

[Pl. 18, fig. 45, eyes. Legs arranged 4. 1. 2. 3. Taken in March. *Supplement.*]

## Genus CYLLOPODIA. Mihi.

Characters. *Cheliceres small; maxillæ short, inclined over the lip; lip wider than long, triangular; eyes six, sub-equal, two very small, placed near together in the middle, two larger far apart placed above, and two placed each on a tubercle on the side; feet, fourth pair longest, then the first, the third shortest.*

*Habits.* Araneides sedentary, making a cocoon.

*Remarks.* This anomalous spider appears to be related to Epeira. I found it in the attitude of one, suspended from a thread or web which I would have examined carefully, had I not taken it as a new species of that sub-genus. It has certainly six eyes only; its cephalothorax is flattened in the middle, being deeply excavated behind; the last joint of its palpi are terminated with a small nail; its abdomen is gibbous and rugose, covering anteriorly a great part of the cephalothorax, with four mammulæ and a cauda; the two anterior pair of legs are directed forward, and the other two in the opposite direction, so that the sternum has a vacant place in the middle.

The trophi are nearly those of Epeira, but approach Theridium. The cheliceres are very small, but capable of reciprocal motion.

**Cyllopodia cavata.**

Pl. 12, fig. 3.

*Description.* Piceous; cephalothorax deeply excavated at base for the reception of the abdomen; abdomen varied with white dots and lines, five tubercles covered with tufts of scales on each side above, the second from the base hornlike; feet, 4. 1. 2. 3. two anterior pair directed forward, the other two turned backward, leaving a vacant space on the sternum.

*Observations.* This was found on a twig near an Epeira.

*Habitat.* Alabama, October.

[Pl. 18, fig. 80, eyes. *Supplement.*]

[♀, length 8.9 mm.; cephalothorax 3.1 mm.; legs 3, 2.4, 2, 3.3.
Pl. 20, fig. 21, ♂, ♀. Palpus of ♂ and a hind tarsus of ♀ showing calamistrum.

This no doubt belongs in the genus *Uptiotes* Walck., or *Mithras* Koch. It really has eight eyes, the front lateral pair being very small and colorless. Found in pine woods among dead branches, which it much resembles in color. The web and habits are described by Prof. B. G. Wilder, in Proceedings of the American Association for the Advancement of Science, 1873, p. 265.

Beverly, Mass., Aug. 28; Readville, Mass., Sept. 17, ♂ and ♀, in webs on dead trees; two females making their webs at sunset. Providence, R. I.; Peak's Island, Me.; Ithaca, N. Y., B. G. Wilder. J. H. E.] .

## Sub-genus PRODIDOMUS. Mihi.

*Characters.* *Eyes eight, placed near together, four in* [467] *front, making a straight row, two on each side, forming a curve with the external ones of the first row, and leaving a space above; external ones sub-oval, two middle ones round and black; maxillæ triangular, wide at base, pointed at tip; cheliceres very large, fangs long and bent; feet* $\overline{4.\ 1.}\ 2.\ 3.$

*Observations.* This new sub-genus shows some of the characters of Clubiona and of Theridion. I hope some future naturalist will give its history and its location in the natural arrangement. I know nothing of its habits.

### Prodidomus rufus.

Pl. 12, fig. 3.

*Description.* Rufous; abdomen deeper above, venter pale, four nipples; feet, $\overline{4.\ 1.}\ 2.\ 3.$

*Habitat.* Alabama, in dark cellars.

[Pl. 18, fig. 9, eyes. The three external eyes are oval, shining white. Taken August 10th, in the recess of a large box in a dark cellar, hiding itself in holes. *Supplement.*]

## Genus EPEIRA. Walck.

*Characters.* *Cheliceres short; maxillæ parallel, short, wide at base, truncated at tip; lip wide, sub-triangular; eyes eight, four in the middle placed in the form of a square, two on each side placed near each other diagonally on a common eminence;*

*feet, commonly the first and second longest, the third being the shortest.*

*Habits.* Araneides sedentary, forming a web composed of spiral threads crossed by other threads departing from the centre, often dwelling in a tent constructed above the web. Cocoon of various form.

*Remarks.* I endeavored to arrange the numerous species of this sub-genus according to the method of Walckenaer; but the characters of Epeira are not very liable to vary, except by the form of its body. The middle eyes offer some variations, it is true, and the lateral ones are sometimes placed lower than in others; but I could not avail myself of these characters to establish natural subdivisions.

The spiders of this sub-genus are known to every observer of nature. Their habits, and particularly their webs, are familiar [468] to every one. Their history enters into the history of man. If it be not a fiction, it was a spider of this section which, by making its web at the entrance of the cave concealing Mahomet, saved the life of the impostor. The description of Ovid is sufficient to show that the ill-fated *Arachne* was transformed into an Epeira by the Goddess Pallas, or rather by the observant poet of Sulmo. [Ovid's Metamorph., VI, p. 141.]

Tribe I. OVATÆ INERMES. *Body without spines, generally large.*

### 1. Epeira riparia.

*E. fasciata?* R. A., IV, p. 219.

Pl. 12, fig. 5.

*Description.* Black, cephalothorax covered with silvery white hairs, abdomen varied with bright yellow spots and dots. Thighs usually bright rufous at base, except the first pair. Of a large size, seldom small.

*Observations.* This remarkable species usually dwells on the margin of waters, where it makes a web of strong threads, in which large Libellulæ are often caught. The abdomen of the female is flat in the early part of the season, and it is not till August, that being distended with eggs it assumes the oviform shape. Its cocoon is conical, as large as a small plum, like a pear hanging down. Whenever opened it was found full of young spiders instead of eggs. Is it viviparous?

*Habitat.* The United States.

[Pl. 19, fig. 121. Abdomen beneath. *Supplement.*]

[♀, length 18 mm.; cephalothorax 6 mm.; legs 25, 24.5, 15†2; 24.5.
♂ " 5.5 mm.; " 2.5 mm.; legs 13, 13, 6.5, 10.
Pl. 21, fig. 1, ♂ enlarged twice, and palpus of ♂.
Probably *Argiope aurantia* Lucas, Ann. Soc. Ent. France, Vol. ii, 1833.
Beverly, Mass., July 14; young in round webs; September, ♂, in irregular webs near the round webs of ♀; December, cocoons of young on bushes. Racine, Wis.; Ohio; Texas; Mayport, Fla.; Panama, Wm. Holden. J. H. E.]

**2. Epeira fasciata?** R. A. IV, 249.

Pl. 12, fig. 8.

*Description.* Covered with silvery white hairs; abdomen with about fifteen transverse, nearly interrupted black bands, and several yellow marks between; feet rufous with black bands, anterior thighs black. [469.]

*Observations.* This spider should be dedicated to the greatest Entomologist of this age, (Latreille) if it proves not to be the *fasciata.* No doubt it is related to *E. fasciata* of Europe. (R. A., IV, 249.) It is quite rare in the Southern States, but common in New England, particularly in Maine. It abounds in meadows, near the ground, where it makes its web. An immaculate species was found in North Carolina, which may be referred to this, as its abdomen was not distended with eggs, and the bands may become apparent when it is full. It was surrounded with several males four or five times smaller.

*Habitat.* United States.

[The presence of the eggs in the abdomen always creates a change in colors; legs arranged 1. 2. 4. 3. *Supplement.*]

[♀, length 21 mm.; cephalothorax 7 mm.; legs 31.5, 30, 18.2, 29.

♀ " 16 mm.; " 5.6 mm.; legs 25.4, 24.8, 15, 23.

♂ " 3.8 mm.; " 1.8 mm.; legs 8, 7.8, 3.7, 7.

Pl. 21, fig. 2, ♂, enlarged twice, and palpus.

Beverly, Mass., Aug. 13, ♂, and young; Sept. 8, ♀, in round webs near water. (Ohio, ♀, Wm. Holden.) J. H. E.]

### 3. Epeira vulgaris.

Pl. 12, fig. 6.

*Description.* Pale gray, abdomen piceous, with various winding white marks, a middle one in the form of a cross; feet with piceous rings.

*Observations.* This spider is well known even to those who are not attentive observers of nature. Every one has noticed its regular geometrical web, which is frequently placed near the windows of our houses. It is subject to such variations in color and marking that it is quite difficult to distinguish several other species from varieties of this species. I have once found seventeen varieties of spiders enclosed in the nest of a Sphex, called *dirt-dauber* in the Southern States, and each could be referred to this species, though they all differed more or less from each other. This species seems domesticated, being seldom found far from our gardens. The reason probably is, that it is more secure there from its enemy, the Sphex.

*Habitat.* South Carolina.

[Pl. 18, fig. 88, eyes; fig. 103, trophi. Legs arranged 1, 2, 4, 3. *Supplement.*]

[♀, length 15 mm.; cephalothorax 6 mm.; legs 2.4, 21, 12.4, 17.8.

♂ " 8.4 mm.; " 4.2 mm.; legs 28, 23.2, 12.6, 17.5.

Pl. 21, fig. 4, palpus of ♂.

Boston, Mass.; Portland, Me.; Providence, R. I.; Albany, N. Y.; Appleton, Wis. J. H. E.]

### 4. Epeira domiciliorum.

Pl. 12, fig. 7.

*Description.* Gray or brownish, covered with coarse white hairs; cephalothorax with a blackish band near the edge; [470]

abdomen with many markings of black and dusky surrounding a spot in the form of a cross; thighs rufous at base, tipped with a blackish ring, other joints with dusky rings; abdomen underneath with a large black spot, near the centre of which are two white dots. A large species.

*Observations.* This spider is often found in dark places, and even in dark apartments not much frequented, where it makes its web. The female is supplied with a hook over the vulva as in *E. diadema.* See Régne Animale, IV, p. 218. It makes a cocoon of yellow silk in the shape of a button, lenticular, and attached to a solid body.

*Habitat.* Alabama. July, September.

[Pl. 19, fig. 123; lateral and ventral view of the abdomen. Legs arranged 1. 2. 3. 4. *Supplement.*]

[♀, length 10.8 mm.; cephalothorax 3.9 mm.; legs 15, 13.9, 8.2, 14.
Salem, Mass., Sept. 6; on fences; Cambridge, Mass., Aug., ♂; Providence, R. I.; Hartford, Conn. J. H. E.]

**5. Epeira septima.**

Pl. 12, fig. 9.

*Description.* Rufous, spotless, hairy above; abdomen with two impressed dots above, and with two angular lines yellowish beneath, thus \/, blackish in the centre; feet varied with yellowish and deep rufous.

*Observations.* This large species is not rare, and is found more commonly in the Autumn. When caught it spins, probably for defence, a large quantity of beautiful white silk which it draws out by pressing its posterior feet against the abdomen.

*Habitat.* North Carolina, Alabama.

[The hair is yellow. Legs arranged 1. 2. 4. 3. It makes a very high web. Taken in September. *Supplement.*]

**6. Epeira insularis.**

Pl. 12, fig. 10.

*Description.* Cephalothorax rufous; abdomen yellow, with many waving purplish markings; thighs and proximate joints orange with rufous rings, terminal joints white, with black rings. A large species.

*Observations.* This is no doubt related to *E. trifolium, aureola,* and *obesa,* but the rufous rings which are found on all the thighs of this, are wanting in those. Like *E. domiciliorum,* the female has a small hook above the vulva. [**471.**]

*Habitat.* Found on an island of the Tennessee, Oct. 13th, after some frost.

[♀, length 16.2 mm.; cephalothorax 6.3 mm.; legs 20.4, 19.3, 18.2, 19.4.
♂ " 6.7 mm.; " 3.7 mm.; legs 13, 12.8, 6.7, 10.4.
Pl. 21, fig. 3, 3*a*, 3*b*, palpus of ♂.
Beverly, Mass., Aug. 13; in tents on bushes five feet from the ground.
Cambridge, Mass., Sept. 21; ♀ in web, 2 ♂ near in an irregular web.
(Ohio, Wm. Holden.) J. H. E.]

### 7. Epeira obesa.

Pl. 12, fig. 11.

*Description.* Testaceous; abdomen with obscure marks, nearly orbicular, feet with joints tipped with rufous.

*Observations.* This species was found after the first frosts, its abdomen still filled with eggs.

*Habitat.* Maine.

[Pl. 18, fig. 46, eyes. Legs arranged 1. 4. 2. 3. *Supplement.*]

### 8. Epeira trifolium.

Pl. 13, fig. 1.

*Description.* Cephalothorax pale, with three longitudinal blackish bands; abdomen purplish, with many spots and two undulated bands white; joints of feet tipped with black, posterior thighs with one black ring near the middle.

*Observations.* This elegant species is one of those which it is very difficult to distinguish from others. It was found in houses and near dwellings.

*Habitat.* Maine.

[Legs arranged 1. 2. 4. 3. *Supplement.*]

[♀, length 14 mm.; cephalothorax 6.2 mm.; legs 20.3. 18.6, 12.5, 17.5.

This is very near *Epeira quadrata* of Europe, if not the same species. The females are often much larger than Hentz's figure, and vary in color from dark purplish brown to light yellow.

Eastern Massachusetts, September and October, in tents formed by spinning several leaves together near their webs.   J. H. E.]

### 9. Epeira aureola.

Pl. 13, fig. 2.

*Description.* Pale testaceous; cephalothorax rufous; abdomen orange color, with white dots of various sizes; joints of feet tipped with rufous, posterior thighs with one rufous ring near the middle.

*Observations.* This species certainly differs much in markings from *E. trifolium*, and yet, being found after the first frost, it may have changed by the cold, and prove a mere variety of that species. For the present, however, I consider it as distinct.

*Habitat.* Maine.

[Legs arranged 1. 2. 4. 3.   *Supplement.*]

### 10. Epeira labyrinthea.

Pl. 13, fig. 3.

*Description.* Reddish brown, abdomen varied with paler [472] spots, and a scolloped white line above, a white longitudinal line edged with black and two white dots near the apex beneath; feet rufous, yellowish towards the end; male the same, with hairy legs.

*Observations.* This very distinct and common species is of middling size, seldom larger than the drawing. Its web is very compound, for it unites together that of a Theridium, partly that of an Agelena, and that of an Epeira. The web peculiar to this sub-genus is in front, then usually a tube like that of Agelena leads from this to one made of crossed threads like that of Theridium, at the upper part of which is constructed a tent covered with dried leaves in the manner of shingles, under which it remains sheltered during the day. The cocoon is in the shape of a button or flattened cone, sometimes brownish above and pale gray beneath ; as many as five have been found in a string, one above the other. The young when just hatched resemble the mother. . The first time I found this spider, I also found the first Mimetus, which had invaded the web of one of these and taken its place, so that for a period I thought this species a transition to that sub-genus. But this has all the characters of Epeira.

*Habitat.* North Carolina, Alabama.

[Pl. 18, fig. 25, eyes; fig. 93, trophi, wanting the palpus; pl. 19, fig. 124, web and cocoons; fig. 133, web and tent. Legs arranged 1. 2. 4. 3. Very common in damp woods. Taken in September, October and December. *Supplement.*]

### 11. Epeira prompta.

Pl. 13, fig. 4.

*Description.* Pale bluish ; abdomen with two indented lines and several spots black ; base of the thighs rufous, a black ring between this and the tip, and black rings on the other joints.

*Observations.* This very distinct species is very active after sunset, running with great speed, and leaping like an Attus. It is motionless during the day. A small specimen of this species was found in Alabama, with its abdomen black underneath, having a central whitish spot.

*Habitat.* Massachusetts. Alabama. [473.]

[Pl. 18, fig. 47, eyes. *a*, specimen from Massachusetts. *b*, specimen from Alabama. Legs arranged 1. 2. 4. 3. Taken in Massachusetts in June. *Supplement.*]·

### 12. Epeira strix.

Pl. 13, fig. 5.

*Description.* Rufous ; abdomen yellow, with a scolloped blackish band on each side above, and about six black dots, a broad black spot underneath, with a yellow lunule on each side ; feet, with joints terminated by a black band.

*Observations.* The male and female were found very frequently near streams, where they make perpendicular webs. This spider during the day remains strictly, concealed near its web, in a dwelling which it constructs with leaves drawn together in the manner of a tube by means of threads.

*Habitat.* Pennsylvania ; Alabama.

[♀, length 10.8 mm.; cephalothorax 5 mm.; legs 14, 13, 8.2, 11.6.
♂, " 10 mm.; " 4.8 mm.; legs 15.8, 13.4, 9, 12.
Eastport, Me., Aug.; Portland, Me., Aug.; Noank, Ct., Aug. 20, ♂ and ♀ ; Boston and vicinity. Males mature about Sept. 1. J. H. E.]

**13. Epeira Thaddeus.**

Pl. 13, fig. 6.

*Description.* Cephalothorax rufous ; abdomen green, yellowish towards the base, with a black band on each side of the abdomen, piceous underneath, with a yellow spot in the centre ; feet orange, varied with rufous and blackish. A somewhat large species.

*Observations.* This species, which is sometimes whitish on the abdomen, is nevertheless very readily recognized. I have seen some specimens larger than the delineation. Its dwelling-place is really beautiful ; it is placed above its web, and made of the finest white silk, shining with a satin lustre ; its shape is that of an inverted thimble, and it is usually placed under a leaf bent together for the purpose of affording shelter and security.

*Habitat.* Alabama. September, October.

[A specimen was found in October, larger than the drawing, Pl. 13, fig. 6, but the abdomen was nearly white instead of green. *Supplement.*]

**14. Epeira hebes.**

Pl. 13, fig. 7.

*Description.* Brown, abdomen with several forked lines, and two spots black.

*Observations.* This would appear to be an obscure species, and not easily distinguished from *E. vulgaris* ; but, being first [474] described in South Carolina, then seen in North Carolina several years afterwards, I consider it as a distinct species. It is perfectly inactive in the daytime, living chiefly on coleopterous insects, which it binds up in a few minutes with a strong web of silk.

*Habitat.* Southern Atlantic States.

[Pl. 18, fig. 81, eyes. Legs arranged 1. 2. 4. 3. *Supplement.*]

### 15. Epeira maura.

Pl. 13, fig. 8.

*Description.* Rufous ; abdomen oval, black, highly glossy, with yellow spots, underneath blackish, spotted with yellowish ; feet varied with black rings. A middle-sized species.

*Observations.* This very distinct species was usually found in the vicinity of streams of water.

*Habitat.* Alabama. April, May, September.

[Legs arranged 1. 2. 4. 3. *Supplement.*]

### 16. Epeira nivea.

Pl. 13, fig. 9.

*Description.* White above and beneath ; abdomen nearly orbicular, with an oval blackish spot on the disc.

*Observations.* This spotless species is remarkable for its pale color, and in that respect approaches *E. alba*, but it differs from it by the form of its abdomen, and by its more slender legs.

*Habitat.* Alabama. July.

[Pl. 18, fig. 48, eyes. Legs arranged 1. 2. 4. 3. *Supplement.*]

### 17. Epeira? hamata.

Pl. 13, fig. 10.

*Description.* Whitish ; abdomen with a blackish band broad at base, and terminating in a point before the apex ; feet (in

the male) varied with blackish, with a few long hairs, second
pair with the antepenultimate joint crooked, having one bristle
longer than the rest.

*Observations.* The characters of this somewhat depart from
Epeira. The web has not been observed, and the female is
unknown. Could it be the male of *E. nivea?*

*Habitat.* Alabama. August. [475.]

[Pl. 18, fig. 49, eyes. *Supplement.*]

**18. Epeira pratensis.**

Pl. 13, fig. 11.

*Description.* Yellow; abdomen yellowish rufous, with two
rows of black dots, approaching each other towards the apex.

*Observations.* This spider, found in a field, was seen only
once.

*Habitat.* Massachusetts.

[Legs and cephalothorax immaculate, yellow; legs arranged
1. 2. 4. 3. Taken in July. *Supplement.*]

[♀, length 7 mm.; cephalothorax 2.6 mm.; legs 8.2, 7.5, 5, 7.2.
Beverly, Mass., July 15, in a web on a bridge; Chelsea marshes, July, ♂
and ♀. J. H. E.]

**19. Epeira placida.**

Pl. 13, fig. 12.

*Description.* Yellowish or pale rufous; cephalothorax with
an obscure band, and darkish edge; abdomen varied with
whitish, brownish lines, and an angular piceous band; feet
hairy. A small species.

*Observations.* This may be distinguished from *E. spiculata*
by its marking, but particularly by the lower middle eyes,
which are farther from each other than the upper ones. It
makes a perpendicular web.

*Habitat.* Alabama. April.

[Pl. 18, fig. 30, eyes. Legs arranged 1. 2. 4. 3. *Supplement.*]

[♀, length 3.4 mm.; cephalothorax 1.4 mm.; legs 5, 4.8, 2.8, 4.4.

♂, "   2.4 mm.;   "   1.2 mm.;  "  4.8, 4.7, 2.3, 3.6.

Beverly, Mass., May 10, young; July 1, ♂ and ♀.; New Haven, Ct., Oct., young; Providence, R. I.; Mt. Tom, Mass.  J. II. E.]

### 20. Epeira spiculata.

Pl. 13, fig. 13.

*Description.* Pale or yellowish; cephalothorax with a narrow blackish band; abdomen whitish, with a barbed purplish black band; feet hairy, with a shade of greenish. A very small species.

*Observations.* This species is very common in the woods, making a perpendicular web.

*Habitat.* Alabama. September, October.

[Pl. 18, fig. 31, eyes; fig. 94, trophi. Legs arranged 1. 2. 4. 3. *Supplement.*]

### 21. Epeira foliata.

Pl. 13, fig. 14.

*Description.* Pale brown; abdomen ovate terminating in a joint, with waved black lines, two external almost meeting at the apex, two internal meeting before or near the middle. [476.]

*Observations.* This spider is not unfrequently found on weeds and bushes. When at rest it gathers some leaves together as a tent. It moves with amazing rapidity. It is quite distinct from *E. hebes.*

*Habitat.* Alabama. June, July.

[Pl. 18, fig. 50, eyes. Legs arranged 1. 2. 4. 3. *Supplement.*]

### 22. Epeira sanguinalis.

Pl. 13, fig. 15.

*Description.* Greenish yellow; abdomen above crimson with about three central spots, and two opposed bands scalloped towards the median line, white. A small species.

*Observations.* This singularly marked spider was found in my cocoonery, suspended from a thread.

*Habitat.* Alabama. July.

[Pl. 18, fig. 62, eyes. Legs arranged 1. 2. 4. 3. *Supplement.*]

### 23. Epeira bombycinaria.

Pl. 13, fig. 16.

*Description.* Cephalothorax rufous; abdomen grayish above and beneath, above with two white spots near the base, two little dots near the middle, and two opposed scalloped lines blackish; feet rufous, varied with black.

*Observations.* This spider was found making its web, and abiding in my cocoonery.

*Habitat.* Alabama.

[Legs arranged 1. 2. 4. 3. Taken in August. *Supplement.*]

### 24. Epeira displicata.

Pl. 13, fig. 17.

*Description.* Yellowish or whitish; cephalothorax with two upper middle eyes much larger than the rest; abdomen with four impressed dots before the middle, and six black dots near the apex, each dot surrounded by a pale ring. A small species.

*Observations.* This spider is very common in the spring in low bushes and grass.

*Habitat.* Alabama. May, October. [477.]

[Pl. 18, fig. 51, eyes. Legs arranged 1. 2. 4. 3. *Supplement.*]

[♀, length 7.6 mm.; cephalothorax 2.8 mm.; legs 7.2, 6.8, 4.8, 7.

♂, " 4.5 mm.; " 2.1 mm.; " 6.5, 6, 4.3, 6.2.

Pl. 21, fig. 5, palpus of ♂.

Small specimens often have the abdomen bright crimson, which becomes brighter after the spider has been a short time in alcohol.

Beverly, Mass., June 4, in web in woods; Saugus, Mass., June 12, ♂ and ♀; Dedham, Mass., Nov. 9, on fence without web; Providence, R. I. (Marietta, Ohio, ♂. W. Holden.) J. H. E.]

### 25. Epeira rubens.

Pl. 13, fig. 18.

*Description.* Red; cephalothorax with the region of the eyes black; abdomen with four impressed dots placed somewhat anteriorly; feet black, except the base of the thighs, which is red like the palpi and the rest of the body. A small species, about the size of the drawing.

*Observations.* This species is not rare, both sexes having been found on perpendicular webs. It will be readily distinguished from *L. coccinea*, by the form of its abdomen, which is nearly orbicular, by its legs, which are short, etc.

*Habitat.* Alabama. June, October.

[A specimen was found corresponding in every respect, except in having two obscure spots near the end of the abdomen. Taken in Alabama, Oct. 13th, on grass, beginning to throw threads from one blade to another; another was found in June, slightly larger than is indicated by the side lines in Fig. 18. Legs arranged 1. 2̄. 4̄. 3. *Supplement.*]

Tribe II. OVATA INCLINATA. *Body sub-cylindrical, web horizontal.*

### 26. Epeira hortorum.

Pl. 18, fig. 19.

*Description.* Tender green; abdomen above silvery white with branching blackish lines, beneath green with yellowish lines and an orange spot.

*Observations.* This truly elegant and common inhabitant of gardens and fields, makes an horizontal web on bushes generally sheltered from strong winds, which would destroy its slender fabric. Its web is extremely regular, and is a fair sample of the skill of the genus Epeira.

*Habitat.* All the United States.

[Pl. 18, fig. 10, eyes. Legs arranged 1. 2. 4. 3. *Supplement.*]

[♀, length 6.5 mm.; cephalothorax 2.4 mm.; legs 16.4, 12, 5.8, 10.

St. John's River, Florida, E. Palmer; Newbern, North Carolina, R. Berry; Florida, E. Palmer. J. H. E.]

### 27. Epeira gibberosa.

Pl. 13, fig. 20.

*Description.* Pale yellowish green, body covered with round yellow dots; cephalothorax elevated in the middle; abdomen yellowish above, with three longitudinal and several diagonal lines purplish black, and three or four transverse orange lines, underneath greenish with blackish lines and small whitish spots; feet hairy, 1. 4. 2. 3. in all specimens. A species of moderate size. [478.]

*Observations.* This very distinctly marked species is by no means common, dwelling in low grassy places, and making sometimes perpendicular, sometimes horizontal webs. Except in the respective length of its feet, it is related to *E. hortorum.*

*Habitat.* Alabama. September, October.

[Pl. 18, fig. 11, eyes. *Supplement.*]

[♀, length 3.6 mm.; thorax 1.7 mm.; legs 7, 6.4, 4.8, 7.

♂, " 3 mm.; thorax 1.5 mm.; legs 6.7, 5.3, 3.4, 5.8.

Pl. 21, fig. 6, palpus of ♂.

Holyoke, Mass, July 4; ♂ and ♀; Groveland, Mass., July 16, ♂ and ♀; Beverly, Mass., Aug. 13. J. H. E.]

Tribe III. ELONGATÆ CYLINDRICÆ. *Body elongated, cylindrical.*

### 28. Epeira directa.

Pl. 13, fig. 21.

*Description.* Pale testaceous; abdomen oblong, with three or four parallel rufous lines on each side of a central one which alone reaches the base.

*Observations.* This spider is found generally near water, where it makes a perpendicular web on low bushes. When approached, it drops down and remains motionless where it falls. Its body is very soft. The same was found in Alabama, differing only in having about four minute blackish dots on the second line from the central one on the abdomen.

*Habitat.* South Carolina, Alabama.

[A male was also found April 25, with black dots all over the legs, except on the thighs, and also with six black dots on each side of the abdomen; but evidently the same species; legs arranged 1. 2. 4. 3. It is nocturnal in its habits. *Supplement.*]

### 29. Epeira rubella.

Pl. 13, fig. 22.

*Description.* Bright rufous; abdomen oblong, with a white longitudinal band; feet slender; a very small species.

*Observations.* This little spider was found on a web which I think was perpendicular. It may prove to be the young of *E. directa,* but it is not probable.

*Habitat.* Alabama. August.

[Legs arranged 1. 2. 4. 3. *Supplement.*]

### 30. Epeira sutrix.

Pl. 13, fig. 23.

*Description.* Whitish; covered with short silvery hair, varied with brownish abbreviated lines, and bands on the feet. [479.]

*Observations.* This spider, found on the sea shore, has the talent of adding to the strength of its web, in places where the wind ever moves it, by adding to it strong white threads in a zigzag manner, just as a seamstress darns stockings. This is usually done between two of the concentric lines, and also in a circular way in the middle of the web.

*Habitat.* South Carolina; Pennsylvania.

[Pl. 18, fig. 70, eyes; fig. 95, trophi. Legs arranged 1. 2. 4. 3. *Supplement.*]

---

[Continued from Vol. v. p. 479.]

Tribe IV. MAMMULOSÆ. *Abdomen with various projections.*

### 31. Epeira? pentagona.

Pl. 14, fig. 1.

*Description.* Varied with yellowish and whitish, marked

with black spots; cephalothorax elongated, external eyes separated; abdomen with four tubercles; feet 1. 2. 4. 3, the first and second much the longest.

*Observations.* This may constitute the type of a new [19] subgenus, as it makes a horizontal web. For the present it may be placed in my tribe of the *Ovatæ inclinatæ.*

*Habitat.* Alabama.

[Pl. 18, fig. 20, eyes. The first and second pairs of legs are much longer than the fourth. *Supplement.*]

### 32. Epeira verrucosa.

Pl. 14, fig. 2.

*Description.* Purplish brown or black, varied with yellowish; body very soft; abdomen with a large triangular spot, glossy yellow or whitish; six or eight tubercles more or less prominent at the apex, sometimes with a white dot near their root; base with a tubercle on each side; feet annulated with brownish.

*Observations.* This species is very distinct, and not rare, usually preferring shady places, and the vicinity of waters.

*Habitat.* North Carolina; Alabama. July; October.

[Legs arranged 1. 2. 4. 3. *Supplement.*]

[Ohio, ♂ and ♀, Wm. Holden. J. H. E.]

### 33. Epeira scutulata.

Pl. 14, fig. 3.

*Description.* Pale yellowish; abdomen in the form of an escutcheon; varied with orange, and yellow spots; two rows of black dots approximating towards the apex and another row at base. A small species.

*Observations.* This species makes the common web, and remains above in a web of crossed threads like that of Theridium.

*Habitat.* Alabama. October.

[Legs arranged 1. 2. 4. 3. *Supplement.*]

### 34. Epeira infumata.

Pl. 14, fig. 4.

*Description.* Dusky gray; abdomen with two lunate spots and several abbreviated lines; bi-tuberculated at base; thighs tipped with black.

*Observations.* This spider is readily distinguished from any other by its form and color. The lateral eyes are placed nearer the edge of the cephalothorax than the middle ones, and this character is possessed by *E. stellata (?), prompta, hebes,* and *spinea.*

*Habitat.* North Carolina; Alabama. [20.]

[Pl. 18, fig. 52, eyes. Legs arranged 1. 2. 4. 3. Taken June 30. *Supplement.*]

### 35. Epeira heptagon.

Pl. 14, figs. 5, 6.

*Description.* Blackish; abdomen with an angular line on each side near the base, and several small irregular spots white or yellowish; seven or nine projections placed round the edge, the two interior ones largest, producing in their intervals seven or nine sides; two angular white spots underneath; male of a much lighter color, but with the same general markings.

*Observations.* This was first found in the clay tube of a *Sphex cyanea,* along with *Epeira alba* and thirty-eight specimens of *Theridion lineatum.* It makes a perpendicular web, and drops from it when threatened with the slightest danger.

*Habitat.* North Carolina; Alabama.

[Pl. 18, fig. 53, eyes; fig. 72, outline of cephalothorax. Body black, or sometimes rufous; the legs are black, with pale rings; a male was found in Alabama with rufous hairy legs. Legs arranged 1. 2. 4. 3. Taken in July. *Supplement.*]

### 36. Epeira alba.

Pl. 14, fig. 7.

*Description.* Cream white; abdomen with a tuberculated

projection each side, anteriorly, a blackish spot between these, and two dots on the disc; legs with pale blackish rings.

*Observations.* Found in the clay nest of *Sphex cyanea.* It must be very rare.

*Habitat.* North Carolina.

[Pl. 18, fig. 21, eyes. Legs arranged 1. 2. 4. 3. Taken in July. *Supplement.*]

### 37. Epeira cornigera.

Pl. 14, fig. 8.

*Description.* Yellowish; cephalothorax varied with yellow and black, with a bifurcated horn on each side, and many rounded tubercles; abdomen with two tubercles, one on each side anteriorly, and about eight impressed dots on the disc; feet deep yellow, two anterior pair sometimes annulated with piceous. A small species.

*Observations.* This very singular little spider obstinately holds its legs folded up as represented, in the manner of some coleopterous insects.

*Habitat.* Alabama. June, July. [21.]

[Pl. 18, fig. 44, eyes. Legs arranged 1. 2. 4. 3. *Supplement.*]

Tribe V. SPINOSÆ. *Abdomen elongated with spines; feet, fourth pair longest.*

### 38. Epeira spinea.

Pl. 14, fig. 9.

*Description.* Rufous; cephalothorax with a yellowish margin; abdomen with six spines; disc yellow, with black impressed dots; feet 4. 1. 2. 3.

*Observations.* This very singular spider usually makes its web in low bushes, and sometimes places it horizontally. It drops from its web, hanging by a thread, when threatened. Its nipples are borne on a projection, which is an impediment to walking on an even surface. The respective length of the feet depart from the character of Epeira.

*Habitat.* The Atlantic States, but rarely seen in the west.

[♀, length 7.1 mm.; cephalothorax 2.6 mm.; legs 7.4, 6.9, 4.3, 8.

♂, "    4.8 mm.;    "    2 mm.; legs 5, 4.2, 2.6, 4.6.

Pl. 21, fig. 8, palpus of ♂. Pl. 21, fig. 8, ♂.

West Roxbury, Mass., May 25, young, in webs, near the ground; Holyoke, Mass., July 4, ♂, and young ♀; New Haven, Conn., July 22, ♀, in web. When disturbed she leaped to the ground and hid in a dead leaf. Indianapolis, Indiana. (Ohio, ♀; Mayport, Fla., ♀. Wm. Holden.) J. H. E.]

### 39. Epeira rugosa.

Pl. 14, fig. 10, 10*a*, 10*b*.

*Description.* Black; abdomen with ten spines on its edge above, and a large tubercle beneath; disc above with white spots, or white with many black dots and impressed punctures; sides rugose; feet 4. 1. 2. 3. Male very small, rufous; abdomen whitish, with a few blackish maculæ, long and slender without any spine.

*Observations.* This spider, closely related to *E. spinea* in many respects, makes also a web which is usually inclined, sometimes nearly perpendicular. Like that species, when thrown to the ground it moves with great difficulty, on account of the projection of the abdomen downwards. The departure from the characters of Epeira, in the respective length of the legs, shows how wisely nature makes adaptation for each species. Were the fourth pair of legs shorter, the difficulty of motion would be still greater. In this respect particularly, it is related to *Epeira mitrata.* A male was found attached to a female in July, like a pygmy upon a mountain, or rather under a mountain. He was so small that I thought at first it [22] was a parasite preying upon her; one of his palpi was deeply sunk in her vulva, and it was with great difficulty I could separate them. Their copulation in this respect is much like that of dogs.

*Habitat.* The Southern States.

[Pl. 19, fig. 122, lateral view of spider after impregnation.

Legs arranged in some specimens 4. 1. 2. 3, in others 1. 2. 4. 3. Diurnal in its habits. Taken in July and August. *Supplement.*]

[♀, length 9 mm.; cephalothorax 2.5 mm.; legs 6.2, 6, 3.8, 6.6. Indianapolis, Indiana. (Ohio, ♀. Wm. Holden.) J. H. E.]

### 40. Epeira mitrata.

Pl. 14, fig. 11.

*Description.* Pale yellowish or rufous; cephalothorax piceous, margin usually paler; abdomen pale yellow or white, varied with blackish spots and impressed dots above; sides rugose, two spines behind, and two smaller ones a little lower and nearer together, black, with yellow spots beneath and at the sides; feet rufous or piceous, joints paler at base, length 4. 1. 2. 3, or frequently 1. 4. 2. 3. Seldom large.

*Observations.* The abdomen of this singular spider viewed above resembles a bishop's mitre. Its cephalothorax is small and almost concealed by the base of the abdomen. It usually makes its web in low grounds in forests. Its second and third pairs of legs are always shorter than the fourth and first, a character which departs from that of Epeira, and which, with several others, it has in common with *E. rugosa.* It is not very rare.

*Habitat.* North Carolina; Alabama. August, October.

[Pl. 18, fig. 22, eyes. Sometimes there are no transverse bands on the abdomen, and then the black dots, about twenty in number, are more distinct. *Supplement.*]

Tribe VI. STELLATÆ. *Abdomen short and wide, surrounded with short points.*

### 41. Epeira stellata? Bosc.

Pl. 14, fig. 12.

*Description.* Pale brownish, cephalothorax varied with blackish; abdomen rugose, with dull gold colored hair, varied with marks and scalloped bands, and with fifteen conical spines, one before, one behind, and thirteen on the margin; thighs varied with black.

*Observations.* This singular spider always holds its feet drawn up towards the body, and seldom moves in the day-time. [23.] The anterior spine is sometimes much longer and white. Dr. T. W. Harris, of Massachusetts, sent me one specimen with only thirteen spines. The cheliceres are very short and stout in this species.

*Habitat.* The United States.

[Pl. 18, fig. 89, eyes. Legs arranged 1. 2. 4. 3. *Supplement.*]

[♀, length 12 mm.; cephalothorax 4.8 mm.; legs 12.4, 11.2, 6.8, 11. Waverly, Mass., Sept. 26, adult ♀ and young in webs; Swampscott, Mass., Oct. and Nov.; young ♀ on fence posts without webs; Providence, R. I. (Rushville, O., ♀; Ft. Cobb, Indian Terr., ♀. Wm. Holden.) J. H. E.]

## 42. Epeira cancer.

Pl. 14, fig. 13.

*Description.* Black; disc of the abdomen yellowish with black dots, circumference with conical black spines.

*Observations.* This little spider, described or rather delineated by Audubon in his Ornithology, makes perpendicular webs and is not rare in the south, but was never seen in the north.

*Habitat.* South Carolina. Common in South Alabama.

[The abdomen beneath is black, marked with numerous large dots; legs arranged 4. 1. 2. 3. Taken in August. *Supplement.*]

[♀, Florida, Key West. E. Palmer. (Mayport, Fla. Wm. Holden.) J. H. E.]

Tribe VII. CAUDATÆ. *Abdomen much elongated behind, in the females.*

## 43. Epeira caudata.

Pl. 14, fig. 14, 14a, 14b.

*Description.* Female, pale testaceous; cephalothorax piceous; abdomen with a conical projection behind, with many variable markings; joints of the feet tipped with dusky. There are also two tubercles on the disc of the abdomen which become obsolete in many, probably when the body is full of eggs.

Male, rufous; cephalothorax piceous; abdomen with two white dots and a white band above, which are wanting in some, and two white dots underneath; tip of anterior thighs black.

*Observations.* The variations in the form of this spider, and the difference between the sexes, has caused me to describe three species which must be referred to one. It makes a vertical web, on which it attaches its cocoons in a row, sometimes as many as five in number. These are of a brownish color, elliptical, and covered with the remains of the insects [24] which have been devoured by the spider. On examining five of these cocoons attached to the same web, young spiders were found already hatched in the lowest one; those above contained eggs not glued together. Whenever this spider is threatened, it imparts to its web a rapid oscillation, which causes the eye to lose sight of it. This is probably intended to escape destruction from the birds. The male never was seen with a web of his own, but was often found wandering.

*Habitat.* Common throughout the United States.

[Pl. 18, fig. 54, eyes; fig. 96, lip and mandibles. Pl. 19, fig. 116, lateral view of body; fig. 132, web and cocoons. Legs arranged 1. 2. 4. 3., or 1. 2. 4. 3. *Supplement.*]

[♀, length 5.3 mm.; cephalothorax 1.8 mm.; legs 6.5, 5.5, 3.8, 5.4.

♂, " 4.5 mm.; " 2 mm.; " 7, 5.5, 4, 5.4.

Pl. 21, fig. 7, palpus of ♂.

Male much like the female. The male figured by Hentz must belong to a different species.

Peabody, Mass., Apr. 28, half grown young in imperfect webs; Beverly, Mass., May 10, young, in webs, no adults; July 16, in webs with string of rubbish across the centre; Aug. 21, young in web; Peak's Island, Portland, Me.; Providence, R. I.; Albany, N. Y. (Ohio, ♀, Wm. Holden.)

*Epeira conica* Walck., Aptères.

*Cyclosa conica* Menge, Preussiche Spinnen. J. H. E.]

## 44. Epeira Caroli.

Pl. 14, fig. 15.

*Description.* Grayish; cephalothorax black; abdomen much elongated behind, blackish, with the disc grayish, varied with darker lines; feet varied with black, particularly the first and second pairs.

*Observations.* It is not probable that this can be referred to *E. caudata,* though that species varies much in shape.

*Habitat.* Alabama. September.

[The body is piceous beneath; legs arranged 1. 2. 4. 3. It is nocturnal in its habits. *Supplement.*]

### Genus PHILLYRA. Mihi.

Characters. *Cheliceres very short; maxillæ short, parallel, truncated above; lips subtriangular; eyes eight, equal, all borne on tubercles, in two rows of four eyes each; the first nearly straight, placed on the very margin of the cephalothorax, the second arcuated towards the first, so that the external eyes are widely separated from those of the first; feet, the first pair larger and much longer than the rest, the fourth next, than the second, the third being the shortest.*

*Habits.* Araneides sedentary, making a horizontal web formed of spiral threads, crossed by other threads departing from the centre, and abiding on the web with its legs extended in a straight line. Cocoon cylindrical, tapering equally at both ends.

*Remarks.* This new subgenus is probably closely related to Uloborus of Latreille. The position of the eyes, however, is reversed, and the legs are different. In several particulars it is related also to *Tetragnatha.*

The habits of the spider upon which I have established this new subdivision, are analogous to those of *Epeira.* Its web, however, is always horizontal. When threatened, it shakes its web violently and thus escapes the notice of its enemies. The attachment of the mother to her cocoon is really surprising. The web may be taken up with the cocoon attached; and the mother unwilling to leave it, suffers herself to be carried with it, without manifesting the least fear. This may be enclosed in a box, and she will remain by it, apparently contented, if it is not torn from her care.

### 1. Phillyra mammeata.

Pl. 14, fig. 16.

*Description.* Brownish; abdomen with diagonal blackish lines more or less distinct; one tubercle on each side anteriorly; varied with brown and blackish underneath; anterior pair of legs very long and stouter than the rest; the antepenultimate joint with a tuft of blackish bristles above and below near the apex, and usually a pale ring at the base; the other legs varied with whitish and brown.

*Observations.* This spider makes a horizontal web, usually in cavities, among large logs, or in hollow trunks of trees. It shakes its web violently when threatened; and when at rest, being always under it in an inverted position, extends its legs in a parallel line, like Tetragnatha. Its cocoon is made in the shape of a double cone or cylinder, tapering at both ends. It is whitish, with veins of brownish black, and has many small, sharp tubercles. The mother watches it with an incredible perseverance, and cannot be separated from it by any inducement that can be offered. Fear seems to be wholly merged in maternal solicitude; and, as soon as the cocoon is torn from its place, having remained firmly attached to it, she proceeds to secure it with new threads.

*Habitat.* Alabama, in dry places. [26.]

[Pl. 19, fig. 126, cocoon. The lateral eyes of the anterior row are difficult to be seen; the abdomen ends with a nipple-like tail, and is surrounded with six nipples; legs arranged 1. 4. 2. 3. Taken in May, August, September, and October. *Supplement.*]

[♀, length 4 mm.; cephalothorax 1.7 mm.; legs 6.2, 3.5, 2.6, 4.3.

Three light stripes on the cephalothorax are distinct on young specimens, like the young male, *P. riparia* (Fig. 1); in older ones the lateral stripes are wanting, and the central one indistinct. The tarsi of the fourth pair of legs are curved at the base, and furnished with calamistra.

Waverly, Mass., May 18, under a stone; Malden, Mass., II. L. Moody; New Haven, Conn.; Providence, R. I. J. H. E.]

## 2. Phillyra riparia.

Pl. 14, fig. 17.

*Description.* Whitish; cephalothorax with two longitudinal, brownish, narrow bands; abdomen with an interrupted longitudinal line and two lateral curved lines, blackish; one tubercle above near the middle on each side; feet varied with blackish, antepenultimate joint of the anterior pair with two tufts of bristles. Markings of the female pale and indistinct.

*Observations.* This was found on limestone rocks, on the banks of Cypress Creek. It certainly differs from *P. mammeata.*

*Habitat.* North Alabama, in moist places.

[Pl. 18, fig. 61, eyes; fig. 111, fore leg. Legs arranged 1. 4. 2. 3. Taken in March and April. *Supplement.*]

## TETRAGNATHA. Latr.

Characters. *Cheliceres long, serrated, or with prongs; maxillæ parallel, very long, widening at the top, truncated; lip sub-triangular, less than half the length of the maxillæ: palpi long and slender; eyes eight, subequal, in two nearly parallel rows of four each: feet long and slender: first pair longest, then the second, the third being the shortest.*

*Habits.* Araneides sedentary, forming a web composed of spiral threads crossed by other threads departing from the centre, and abiding on the web with their legs extended longitudinally.

*Remarks.* This subgenus is closely related to Epeira, and has nearly the same habits. The species composing it are readily recognized by their long legs extended upon their geometrical webs. They differ greatly in the length of their cheliceres, but in other respects constitute a natural subdivision.

### 1. Tetragnatha grallator.

Pl. 15, figs. 1, 2.

*Description.* FEMALE: Testaceous, abdomen livid above, with a scalloped longitudinal darkish band, darker beneath, with a black longitudinal line and two yellow longitudinal [27] ones. Cheliceres with two rows of teeth, one larger near the apex. MALE: Wholly testaceous or livid. Cheliceres much larger, arched with two rows of teeth and three large prongs; one superior bifurcated at the end.

*Observations.* This spider makes its web on bushes on the margin of springs and rivers. When on a twig it extends all its legs in one straight line. Its web is scarcely ever perpendicular, but inclined, sometimes horizontal.

This may be the *T. elongata* of Bose; but as the name may apply to my *T. laboriosa*, there will be less confusion in using this appellation.

*Habitat.* Pennsylvania, North and South Carolina, Alabama, etc.

[Taken in April and May, one specimen in a dry place in a tree far from water. *Supplement.*]

[Ohio, Wm. Holden. J. H E.]

### 2. Tetragnatha laboriosa.

Pl. 15. fig. 3.

*Description.* Rufo-testaceous: abdomen yellowish with black branching lines above; a black central longitudinal line, and two yellow ones beneath; feet and cheliceres of moderate length; male with the same marking; cheliceres larger but not as elongated as in *T. grallator.*

*Observations.* This spider is found in meadows making the web of an Epeira, and is not found on wet ground more than in dry places. It is very different from *T. grallator*, particularly in the position of its eyes, which in the male and female are placed in two sensibly curved rows; whereas in that species these rows are straight, the upper one almost bent the other way.

*Habitat.* United States.

[Ohio, ♂, ♀. Wm. Holden. J. H. E.]

## LINYPHIA. Latr.

Characters. *Cheliceres moderately long; maxillæ short, parallel, wider and truncated at the top; lip very short, subtriangular; palpi slender; eyes eight, equal, four in the middle, nearly in the form of a square; two each side, placed together on a common elevation; feet slender, the first pair* [28] *longest, then the second and the fourth, the third being the shortest.*

*Habits.* Araneides sedentary, forming a compound web, composed of a horizontal one, which is surmounted by threads irregularly crossed; usually standing in an inverted position under the horizontal web.

*Remarks.* This subgenus is very readily recognized by its singular webs, observable on bushes and weeds, particularly in the morning when covered with dew. There is less ferocity in the spiders of this division than in any other of the family. It is the only subgenus in which the male and female may be seen harmoniously dwelling together.

### 1. Linyphia communis.

Pl. 15, fig. 4.

*Description.* FEMALE: Cephalothorax rufous; abdomen purplish black above, with about five spots on each side, nearly united in the form of two longitudinal bands; farther down are about five smaller white marks; purplish black beneath;

feet greenish brown, short. MALE: Rufous all over; more slender than the female.

*Observations.* This spider, one of the most common in the South, is familiar to every observer of nature. Its perfectly regular webs, when the dew is still on the ground, are seen in great numbers in the fields and gardens. The owner of each web is always found in an inverted position under the horizontal web, which is curved or hollowed downward. The males are very common in the spring, but disappear in the fall. I have observed two males on a web, fighting an obstinate battle; they strove to grasp each other with their cheliceres, and when exhausted by the conflict, they retired at some distance to rest themselves, and presently renewed the combat. I know not how the contest terminated, but I believe it was without bloodshed. During this, the female, who was the lady of the manor, remained very quiet and apparently unconcerned. The ferocious habits of spiders are [29] generally confined to the appropriate sex; for the females are so gentle that I have seen several allow the males to dwell in the same tent with them, the pair living decently together as husband and wife should among Christian people. I saw but once a male alone in a web, and I do not know whether they ever weave one themselves. It is strange that I never saw the cocoon of so common a species.

*Habitat.* United States, though somewhat rare in the North. [Pl. 18, fig. 104, trophi; pl. 19, fig. 118, lateral view. *Supplement.*]

[♀, length 4.2 mm.; cephalothorax 1.6 mm.; legs 6.2, 5.5, 3.6. 5.5.

♂, " 3 mm.; " 1.4 mm.; legs 5.8, 4.5, 3, 4.

Pl. 21, fig. 9, palpus of ♀.

Beverly, Mass., May 10, young in webs; June 23, ♂ and ♀ in a large web; Aug. 13, ♀ only; Peabody, Mass., Aug. 29, young in webs; Sept. 22, in webs in a savin bush. J. H. E.]

## 2. Linyphia marmorata.

Pl. 15, fig. 5.

*Description.* Cephalothorax rufous, with a whitish edge;

abdomen black, with many baids, spots and dots, white with a tinge of yellow; beneath with a few slender white lines and a yellow band each side, interrupted in two places, so as to make about six yellow spots; feet dark green, long.

*Observations.* This is a very large species, and very distinct from *L. communis*, making very large webs, with long threads to secure them.

*Habitat.* Alabama. July, August.

[Pl. 18, fig. 23, eyes. Legs arranged 1. 2. 4. 3. *Supplement.*]

[♀, length 5.3 mm.; cephalothorax 1.8 mm.; legs 12.7, 10, 6.7, 9.5.

♂, " 5.7 mm.; " 2.6 mm.; legs 16.8, 14.6, 8.3, 12.6.

Pl. 21, fig. 10, palpus of ♀.

All my adult females resemble pl. 15, fig. 5. All young specimens are like fig. 6, *L. scripta*.

Swampscott, Mass., March 19, young with imperfect web; April 21, young females in webs among boulders; Beverly, Mass., June 4, ♂ and ♀ in copulation; Peabody, Mass., July 7, ♂ and ♀ in copulation; Aug. 13, few adults, but many young; Sept. 4, young only; Franconia, N. H.; Eastport, Me.; Portland, Me.; Providence, R. I.; New Haven, Conn.; Albany, N. Y. (Ohio, ♂, ♀. Wm. Holden.) J. H. E.]

### 3. Linyphia scripta.

Pl. 15, fig. 6.

*Description.* Cephalothorax blackish purple with a white edge; abdomen white, with curved spots and obsolete marks, purplish; feet pale greenish, long; a small species.

*Observations.* This species may be recognized in the fields by the peculiar form of its web; the horizontal part of which, instead of being curved or hollowed downward, as in *L. communis*, is rounded upwards, so that the spider stands inverted, as it were, under a bowl. It is quite distinct from that species, and from *L. marmorata*.

*Habitat.* Alabama. May–September. [30.]

[Young of *L. marmorata*. J. H. E.]

**4. Linyphia conferta.**

Pl. 15, fig. 7.

*Description.* Cephalothorax yellowish, with an abbreviated blackish line; abdomen whitish, varied at the sides with greenish lines, with a longitudinal dusky band trifurcated towards the base; feet greenish, 1. $\overline{4. 2}$. 3.

*Observations.* This spider makes a web with its curve upward, like an inverted bowl, and remains in its concavity in an inverted position. It was discovered and delineated by my son, Charles A. Hentz.

*Habitat.* Alabama.

[Pl. 19, fig. 115, lateral view. A specimen had the legs of the right side arranged 1. 4. 2. 3; of the left 1. 2. 4. 3. *Supplement.*]

**5. Linyphia coccinea.**

Pl. 15, fig. 8.

*Description.* Crimson or red; last three joints of palpi, area of the eyes, and tip of the tubercle of the abdomen, black; abdomen with a terminal tubercle above the anus; feet yellowish red, 1. $\overline{4. 2}$. 3.

*Observations.* This species is not very rare in North Carolina, but has not been seen in Alabama. It may be readily distinguished from *Epeira rubens* by the form of its abdomen, and other characters.

*Habitat.* North Carolina.

[Pl. 18, fig. 12, eyes. It makes a thread, like *L. communis.* Taken in July. *Supplement.*]

**6. Linyphia? autumnalis.**

Pl. 15, fig. 9.

*Description.* Livid yellow; cephalothorax with a longitudinal band and margin dusky; abdomen with a double row of dots, connected with a longitudinal line; black above; an indented band blackish beneath; feet varied with dusky bands.

*Observations.* This little species, seen only in the North, may possibly be referred to Theridion. It makes a web with threads stretched in all directions, in the corners of walls, dark places, etc., and remains in the middle in an inverted position, like Linyphia.

*Habitat.* Maine and Massachusetts. [31.]

[Pl. 18, fig. 71, eyes; fig. 97, trophi. Only seen late in the autumn. *Supplement.*]

[♀, length 4.2 mm.; cephalothorax 2 mm.; legs 10.8, 9.2, 7, 9.5.
♂, " 4.4 mm.; " 2 mm.; legs 11.8, 10.2, 8, 11.
Pl. 21, fig. 15, palpus of ♂.
Salem, Mass., Jan. 12, ♂ and ♀; May 1; Beverly, Mass., June, E. Burgess; Boston, Mass., Dec. 20, in cellar, behind boxes. J. H. E.]

**10. Linyphia? neophita.**

Pl. 15, fig. 10.

*Description.* Rufous brown; abdomen piceous; small, a male, feet $\overline{1. 4. 2}$. 3.

*Observations.* This small species was found running on the ground, and is placed with doubts in this subdivision. Its abdomen has no projection like *Linyphia coccinea*, and therefore it is not probable that this is the male of that species.

*Habitat.* North Carolina.

[Pl. 18, fig. 13, eyes. Legs arranged 1. $\overline{4. 2}$. 3. Taken in December. *Supplement.*]

**7. Linyphia? costata.**

Pl. 15, fig. 11.

*Description.* Pale yellowish; cephalothorax with a slender blackish line bifurcating towards the eyes; abdomen with a serrated band and diagonal lines, brownish; feet hairy, varied with blackish; thighs with many blackish rings; feet 1. 2. 4. 3. One of the largest species; even larger than the drawing.

*Observations.* This spider may be separated from this sub-division by other naturalists, but the characters derived from its eyes, trophi, and feet, are those of Linyphia. It is only in the form of its web that it departs from it. It makes a large horizontal web, somewhat like that of Agelena, but without a tube ; this is placed under broad leaves, such as those of hickory. The spider remains in an inverted position at one end, where threads are crossed irregularly, like those of Theridion. It does not endeavor to escape like Epeira, but is very easily taken. It probably does not make its cocoon in its web, as none were ever found. The male makes the same kind of web, and resembles the female.

*Habitat.* Alabama, all seasons.

[Pl. 18, fig. 24, eyes. Legs arranged 1. 2. 4. 3. *Supplement.*]

[♀, length 5.4 mm.; cephalothorax 2.2 mm.; legs 10.5, 8.5, 6.5, 8.

♂, " 5.5 mm.; " 2.5 mm.; legs 11.2, 11, 7.5, 8.5.

Pl. 2i, fig. 11, palpus of ♂.

Beverly, May 10, adult female; Aug. 28, ♀ and young; Salem, Oct., ♂ and ♀, on fences without webs ; Eastport, Me.; Portland, Me.; Providence, R. I.; Albany, N. Y. (Ohio, ♂, ♀. Wm. Holden.) J. H. E.]


Mimetus. Mihi.

Characters. *Cheliceres very long, fang small ; maxillæ tapering, inclined over the lip ; lip pointed, triangular ; eyes* [32] *eight ; four in the middle, the two lower ones borne on tubercles and further apart than the two upper ones ; two on each side placed diagonally near each other, on a middle line ; feet long, the first and second pairs much longer than the other two ; first pair bent in the female.*

*Habits.* Araneides wandering, except during the time of the rearing of the young; destructive of other Araneides, and invading their webs. Cocoon oblong, pointed at both ends.

*Remarks.* The parasitic habits of the spiders composing this subgenus, remind the naturalist of the depredations committed

by various Hymenoptera upon many species of insects. The
Mimetus can make a web like that of Theridion, but prefers
prowling in the dark, and taking possession of the industrious
Epeira's threads and home, or the patient Theridion's web,
after murdering the unsuspecting proprietor.

It combines some of the characters of these two subgenera,
but is more closely related to the latter. The extreme length
of its chelicores is quite anomalous.

### 1. Mimetus interfector.

Pl. 15, figs. 12, 13.

*Description.* Pale yellowish ; cephalothorax with a black
band branching towards the eyes ; abdomen with several white
spots near the base, varying in shape and size ; three central
ones at base, often wanting ; a serrated black line on each side,
almost uniting with its fellow at the apex, and several small
transverse ones ; beneath pale, with little black marks, as
above ; feet very long, with long bristles, varied with rufous'
and black ; first and second pairs with the penultimate joint
curved. The male differs slightly from the female ; his legs
being longer, and the penultimate joint of the first and second
pairs nearly straight.

*Observations.* This singular depredator is not rare, and is
usually found in houses. This has enabled me to make many
curious observations on its manners. The first specimen I
found was a female, which had made two cocoons [33] under
a table in my study, near and among the webs of several of
the *Theridion vulgare.* The cocoon differs in shape from that
which is made by the last-named spider. It is oblong, and
tapers equally at both ends, which are secured by many threads
connected with a web like that of Theridion. Like one of this
subgenus, the mother was watching the young, which were
issuing from the lower cocoon. The second specimen observed
was found devouring the eggs of a *Theridion vulgare*, most
probably after having eaten the mother. The next day it had

disappeared. A third one was found dead in the web of a *Theridion vulgare*, which no doubt had killed it. A fourth one was found eating that very same Theridion. This shows that these two species are mortal enemies. I never knew a spider of this species to remain more than two days in the same place. Its habits seem to be nocturnal ; for generally, when discovered in the day-time, it is found in some dark corner, or crevice, with its legs folded in the manner of several species of Epeira.

I sometimes enclosed specimens of this spider-eater with other species of Araneides, in a glass jar, in order to watch its motions. The moment another spider was thrown in, it showed by its attitude that it was conscious of the presence of an enemy. It first moved its first and second pairs of legs up and down ; then slowly approached its victim, and generally killed it. A *Theridion vulgare*, thrown in, manifested great terror ; but after some seeming reflections on fortitude and necessity, it prepared for the mortal combat, and cautiously advanced towards the Mimetus, which moved more slowly. The Theridion, when near, threw out a long thread, on which were several globules of a transparent fluid. This partially succeeded, for the Mimetus was caught by one leg ; and while the Theridion retreated for observation, it was confined, and dragged about for a long time, before it succeeded in freeing itself. The battle presently was renewed, and this time the Theridion was conquered, and eaten.

*Habitat.* Alabama. [34.]

[Pl. 18, fig. 33, eyes ; pl. 19, fig. 127, cocoon. Legs arranged 1. 2. 4. 3. *Supplement.*]

### 2. Mimetus tuberosus.

Pl. 15, fig. 14.

*Description.* Pale or livid green ; cephalothorax with a black mark branching out towards the eyes ; abdomen subconical, with a tubercle near each of the anterior angles on the side, disk brownish, obscure, with pale spots and a serrated black line ; feet hairy, with many black rings ; first and second pairs long, with the penultimate joint bent in the female, nearly straight and shorter in the male.

*Observations.* This is sufficiently distinct from *M. interfector*, in the form of its abdomen, and comparative brevity of its feet, particularly in the male. A female was found changing her skin on the ground. Many have been seen, but none were observed to make any web.

*Habitat.* Alabama. August–October.

[Legs arranged 1. 2. 4. 3. *Supplement.*]

### 3. Mimetus syllepsicus.

Pl. 15, fig. 15.

*Description.* Pale green ; cephalothorax varied with black ; abdomen with a waved line and disk black ; feet and palpi very hairy ; thighs of first and second pair of legs with a black ring near the tip.

*Observations.* This spider was found safely hidden in the tent of an *Epeira labyrinthea*, which it had no doubt first killed. The webs and the cocoon of its victim were uninjured, and it seemed perfectly at home in its new domicil. How long it would have continued to dwell there, and to avail itself of the industry of its predecessor, I cannot tell, as I took it to describe as a new species of Epeira.

*Habitat.* North Carolina.

[Pl. 18, fig. 34, eyes. Legs arranged 1. 2. 4. 3. Taken in October. *Supplement.*]

### Subgenus Thalamia. Mihi.

*Characters.* *Eyes eight, subequal, in two rows on each side of the front part of the cephalothorax, each row curved inward above, and outward below ; maxillæ wider at* [35] *base, inclined over the lip ; chelicères very small ; feet 2. 3. 4. 1.*

*Observations.* Araneides small, forming a tubular dwelling

of silk in the crevices of walls, protected from the sun and rain. This very distinct subgenus has some affinity to Theridion.

### Thalamia parietalis.

Pl. 15, fig. 16.

*Description.* Obscure; cephalothorax pale, with a bifurcated blackish line; abdomen with several dusky small spots; feet slender, 2. 3. 4. 1.

*Observations.* This very active little spider dwells in crevices of walls, in narrow tubes with an orifice, which serve as nets to arrest its prey. It was discovered and delineated by Charles A. Hentz.

*Habitat.* South Alabama.

### Scytodes cameratus.

Pl. 15, fig. 17.

*Description.* Pale testaceous; cephalothorax large, with various curved dusky lines; abdomen varied with dusky dots and lines; feet with dusky rings; 1. 4. 2. 3.

*Observations.* This spider, which dwells in almost total darkness, in closets among rubbish, does not make any visible web, though it obviously belongs to the genus Scytodes of Latreille. It is most commonly found in the folds of old rags or refuse papers, and shows but little activity in its movements, evidently avoiding the light.

*Habitat.* North Alabama.

[The nails of the cheliceres are very minute; lip wide, lanceolate; maxillæ as in Filistota; last joint of palpi more slender than the rest. Taken from April to November. *Supplement.*]

[♀, length 5.6 mm.; cephalothorax 3 mm.; legs 10, 8.2, 6.0, 9.5.

Salem, Mass., June 16, Museum cellar; Aug. 4, in upper room; Boston, Mass., in library, near book cases; Troy, N. Y. J. H. E.

*Scytodes thoracica* Walckenaer, Aptères.]

[Continued from Vol. vi, p. 35.]

## Genus Theridion Walckenaer.

Characters.  *Cheliceres small, cylindrical; maxillæ widest at base, pointed towards the tip, inclined over the lip: lip small, very short, widest at base, subtriangular; eyes eight, equal, four in the middle, nearly forming a parallelogram, two on each side, placed diagonally; feet slender, the first pair longest, the fourth and the second nearly equal, the third being the shortest.*

*Habits.*  Araneides sedentary, forming a web made of threads crossed in all directions.  Cocoon of various shapes.

*Remarks.*  The subgenus Theridion contains many species, a majority of which are very small, and whose webs, made· on the tops of weeds, in bushes, or in retired corners, are familiar to every one.

I could not adopt any one of the families or tribes of Walckenaer, and it is indeed a difficult matter to subdivide this subgenus.  Whether I have succeeded better will be decided by naturalists.

Tribe I.  Geminatæ.  *External eyes approximated.  Spiders usually small.*

### 1. Theridion vulgare.

Pl. 16, fig. 1.

*Description.*  Female.  Varying from a cream white to a livid brown, or plumbeous color; cephalothorax dull rufous, abdomen with various undulated lines; feet with more or less distinct, dark or plumbeous rings, 1. 4. 2. 3.  **[272.]**

Male.  Slender, same colors and markings except on the legs, which are usually rufous, longer, and have their respective length thus, 1. 2. 4. 3.

*Observations.*  This constant and common inhabitant of any dwelling where the broom is not much in use, is very readily recognized, notwithstanding the variations of its colors.  There is, probably, no spider so abundant in the United States, or so well known to the observer of nature. It makes an irregular web in somewhat retired corners, and usually in dark situations, but occasionally also in the open air. The thread of this web is not very strong, but, by its skill and its activity, the spider makes up for the deficiency. The moment it feels by the vibrations that an insect is caught, it proceeds to the spot with caution, if the prisoner is a large one, and with its posterior legs it throws additional threads, with which it binds the victim with surprising rapidity. As soon as the insect is securely bound, it grasps the end of one of its legs with the fangs of its cheliceres, with which it inflicts a wound which stupefies it in a few seconds. If the prey be not too heavy, it lifts it up to the upper part of its web, where it abides; but when the insect is a very large one, it continually throws more threads around it, and from time to time ascends to the top with additional ligatures, which it firmly fixes to the main threads, and which it pulls as tight as possible. These, by their elasticity, gradually tend to lift up the insect, which by its struggles catches and entangles the threads around its limbs, and in course of time it is hoisted to the top, though the process sometimes continues two and even three days. I have seen an *Ateuchus* (*Coprobius*) *volvens* thus lifted up by a little *Theridion vulgare*. In this case the captured victim probably weighed eighty or a hundred times as much as its destroyer. I have also seen large silk-worms hung up by spiders of this species. In that case the weight raised was still greater. Its cocoon is placed also at the top of the web. It is of a brownish color, and made of somewhat loose threads. The eggs are not glued [273] together, and hatch early. Many cocoons are frequently seen on the same web, though usually there is but one at a time with eggs, the others being previously vacated. This proves the immense propagation of this common tenant of our houses.

This spider differs from some other species in its never concealing its home in holes or crevices.

*Habitat.* All the United States.

[♀, length 6.6 mm.; cephalothorax 2.5 mm.; legs 16.8, 11, 8, 12.8.

♂, "     4 mm.;      "      1.8 mm.; legs 13, 9, 6, 8.

Pl. 21, fig. 12. palpus of ♂.

The commonest house spider in New England, near *T. tepidariorum* and *T. sisyphium.*

Salem, Mass.; Providence, R. I.; New Haven, Conn.; Indianapolis, Ind.; Derry, N. H., ♀; Ohio, ♂, ♀; Ann Arbor, Mich.; Racine, Wisc. J. H. E.]

## 2. Theridion serpentinum.

Pl. 16, fig. 2.

*Description.* Rufous; abdomen yellowish glossy, with two winding lines, connected with the sides, black; varied with black underneath; feet 1. 4. 2. 3.

*Observations.* This spider was brought to me from Georgia by Mr. Thomas R. Dutton. A specimen very much resembling this was found in Alabama, but the trophi were those of *Theridion studiosum,* which is not as large a species as this.

*Habitat.* Georgia.

[Lancaster, O., ♂ and ♀. Wm. Holden. J. H. E.]

## 3. Theridion marmoratum.

Pl. 16, fig. 3.

*Description.* Rufous; abdomen with two impressed dots, whitish, varied with black spots and veins, base whitish, piceous underneath, with a few obsolete pale dots; feet 1. 4. 2. 3.

*Observations.* This spider, though closely related to *T. boreale* and *T. serpentinum,* departs from the characters of this subgenus by its trophi, which are those of Epeira, by a strange anomaly. It is not uncommon under stones. The mouth was examined several times, and always presented the character of Epeira. It was always found under stones.

*Habitat.* Alabama. March, June. [274.]

[It has a whitish band on the anterior part of the abdomen over the back part of the cephalothorax. *Supplement.*]

[♀, length 5.2 mm.; cephalothorax 2.6 mm.; legs 8.6. 7.1, 6, 8.4.

♂, " 5.8 mm.; " 2.7 mm.; legs 9, 7.3, 5.7, 8.2.

Pl. 21, fig. 16, palpus of ♂.

Salem, Mass., June 16, under a stone with web; Aug. 26, young; West Roxbury, Mass., June 21, ♂, F. G. Sanborn; Providence, R. I.; New Haven, Conn.; Indianapolis, Ind. J. H. E.]

### 4. Theridion boreale.

Pl. 16, fig. 4.

*Description.* Piceous; abdomen with a whitish band anteriorly, connected with a longitudinal paler one, and with four impressed dots, two more visible than the others. Palpi of the male enormous, as in the plate; feet 1. 4. 2. 3.

*Observations.* This spider makes its web in darker places than *Th. vulgare*, near a crack or crevice, in which it commonly remains concealed. It also makes its web in the crevices of decaying trees. It is not rare.

*Habitat.* The United States.

[The pale lines on the abdomen make an anchor-shaped marking. Taken in Boston, Mass., and Alabama. *Supplement.*]

[♀, length 6.5 mm.; cephalothorax 2.5 mm.; legs 9.5, 7.3, 5.7, 8.4.

♂, " 6 mm.; " 2.6 mm.; legs 10, 7.7, 6.2, 8.2.

Pl. 21, fig. 13, palpus of ♂.

Salem, Mass., ♂ and ♀ under leaves, and in houses in winter; Providence. R. I., May 21, ♂ and ♀ in copulation on a fence; May 24, large ♀ under a stone, with an irregular web; Eastport, Me.; Portland, Me. (Meredith Village, N. H., ♀; Ann Arbor, Mich., ♂, ♀; Racine, Wisc., ♂, ♀; Ohio, ♂, ♀. Wm. Holden.) J. H. E.]

### 5. Theridion studiosum.

Pl. 16, fig. 5.

*Description.* Greenish brown; abdomen above with two scalloped yellowish lines, beneath with some yellow spots; feet with greenish rings; feet 1. 4. 2. 3.

*Observations.* This spider makes its web on bushes, like Linyphia, frequently on a bush of dead leaves; it is horizontal and closely woven, like that of Agelena, and is surmounted by threads crossed in every direction, but there are none underneath. This species has great affinity to Linyphia. It does not remain in an inverted position under the horizontal web, but abides in the middle like other species of Theridion, and, in the same manner as some species of Epeira, it brings together a few leaves as a shelter. When its web is destroyed it does not abandon its cocoon, which is orbicular and whitish, and is placed in the central part of the web. The mother then grasps it with her cheliceres, and defends her progeny while life endures. She also takes care of her young, making a tent like that of social caterpillars for their shelter, and remaining near them till they can protect themselves. This spider is very sedentary, and little inclined to move; always of small stature.

*Habitat.* South Carolina; Alabama. [275.]

### 6. Theridion anglicanum.

Pl. 16, fig. 6.

*Description.* Body, basal joints of palpi, and base of the thighs, red rufous; abdomen without projection or spot; legs 1. 2. 4. 3.

*Observations.* An individual, supposed to be the male of this, was found in September, with the legs 4. 2. 1. 3, and the abdomen black; palpi very large and complicated.

*Habitat.* Alabama, in June.

### 7. Theridion frondeum.

Pl. 16, fig. 7.

*Description.* Bluish white or pale; cephalothorax with a longitudinal black line; abdomen with six small spots, black, united with a central brownish line; pale yellowish beneath; feet with a few black rings, first pair very long, 1. 4. 2. 3.

*Observations.* This distinct species occurred only once, and was found on a weed.

*Habitat.* Alabama. July.

[New Lexington, Ohio, ♀. Wm. Holden. J. H. E.]

### 8. Theridion cruciatum.

Pl. 16, fig. 8.

*Description.* Pale; abdomen obscure piceous, with a scalloped band, whitish; feet with blackish rings, except the third pair. A very small species.

*Observations.* This spider was found in its web, made like that of other species of the subgenus Theridion.

*Habitat.* Alabama. September–October.

[Legs arranged 1. 4. 2. 3, or 1. 2. 4. 3. *Supplement.*]

### 9. Theridion oscitabundum.

Pl. 16, fig. 9.

*Description.* Abdomen yellowish testaceous, with a subobsolete, rufous, abbreviated line; cephalothorax rufous, region of the eyes black; feet 1. $\overline{2.4}$. 3.

*Habitat.* Found in the hollow of a dry leaf; Alabama. [276.]

[It is of the same color beneath as above. Taken March 15. *Supplement.*]

### 10. Theridion sublatum.

Pl. 16, fig. 10.

*Description.* Pale; cephalothorax piceous, pale on the disk; abdomen with piceous markings, sometimes wholly brown above; pale underneath, with a dusky band; feet always pale, 1. 2. 4. 3. A minute species.

*Observations.* This little spider makes its web in the tops of weeds, in the same manner as *Th. morologum.* It is a common species.

*Habitat.* Alabama. May.

[Pl. 18, fig. 113, under surface of abdomen. *Supplement.*]

**11. Theridion funebre.**

Pl. 16, fig. 11.

*Description.* Black; palpi yellow; abdomen with two bent bands at base, and a spot V at the apex yellow; feet yellow, thighs tipped with black, the other joints with rings, and tipped with black, 4. 1. 2. 3. A small species.

*Observations.* This species is very distinct from any other. It was found wandering.

*Habitat.* Alabama. October.

**13. Theridion leoninum.**

Pl. 16, fig. 12.

*Description.* Yellow; cephalothorax varied with black spots and lines; abdomen with two small tubercles anteriorly, blackish towards the base, with two spots and an inverted ⊥ band, blackish; feet hairy, particularly the two anterior pair, varied with black rings, 1. 2. 4. 3. A small species.

*Observations.* This singular little spider makes its web, like other species of Theridion, in dark corners and recesses. The markings on the abdomen make a tolerable resemblance to the face of a lion.

*Habitat.* Alabama. March. [277.]

**13. Theridion morologum.**

Pl. 16, fig. 13.

*Description.* Brownish rufous; cephalothorax with some longitudinal hairs; abdomen with two oblique lines near the base, one near the centre, and one near the apex, all abbreviated; feet 1. 2. 4. 3.

*Observations.* This may prove to be the male of *Th. foliaceum*, but it does not seem probable to me. It was repeatedly found near the ground, making its web on blades of grass. Only males were found, which renders it probable that the female differs in markings.

*Habitat.* Alabama. October 13th. After frost.

[Pl. 19, fig. 128, web. *Supplement.*]

### 14. Theridion foliaceum.

Pl. 16, fig. 14.

*Description.* Pale brownish; cephalothorax rufous; abdomen with about four oblique curved bands on each side; feet pale yellowish, 1. 2. 4. 3. A small species.

*Observations.* This species is usually found making a slender web in the hollow of leaves.

*Habitat.* Alabama. October.

### 15. Theridion roscidum.

Pl. 16, figs. 15, 16.

*Description.* Cephalothorax rufous; abdomen testaceous, with shades of light blue and purple, with four impressed dots and some smaller impressions; testaceous, unspotted underneath; feet pale rufous, 1. 2. 4. 3. Male rufous, with large cheliceres; abdomen piceous, with several rufous spots on a central line. A very small species.

*Observations.* This species makes its web usually in the hollow of large leaves, where the male and female are often found together. The male is usually of a deeper color, and the female is sometimes deeper than represented in the plate. [278.] Its cocoon is somewhat oval, not very regular in shape, of a snow-white color.

*Habitat.* Alabama. April.

[Pl. 19, fig. 129, web with cocoons. *Supplement.*]

[Charlestown, Mass., ♂ and ♀. Wm. Holden. J. H. E.]

### 16. Theridion cancellatum.

Pl. 16, fig. 17.

*Description.* Abdomen ferruginous, with four transverse white bars; thorax fuscous; legs ferruginous, articulated with dusky; legs 1. 2. 4. 3.

*Observations.* Found in a cavity in limestone rock, male

and female in the same web, made on the roof of the cavity, the spiders being in an inverted position.

*Habitat.* Alabama. April.

### 17. Theridion intentum.

Pl. 16, fig. 19.

*Description.* Abdomen yellowish, venter reddish brown, with three transverse orange bands posteriorly, sides black, and also the median line of the back; thorax and legs black; legs 1. 2. 4. 3.

*Observations.* A male and a female were found in the usual web on a bush in a sink-hole on the La Grange Mountain. The male resembled the female in markings.

*Habitat.* Alabama. August and September.

### 18. Theridion blandum.

Pl. 16, fig. 20.

*Description.* Cephalothorax rufous, deeper in a line from the eyes towards the base; abdomen purplish, with an oblong scalloped yellowish spot; feet pale yellowish green, 1. 2. 4. 3, first pair very much longer. A small species.

*Observation.* This spider makes its web in dark corners. Its cocoon is rounded and white. It is closely related to *T. lyra.*

*Habitat.* Alabama. September. [279.]

### 19. Theridion lyra.

Pl. 16, fig. 21.

*Description.* Pale; cephalothorax with a slender black line branching out towards the eyes; abdomen with two basal curved, black lines, and a central branching dusky line, more or less distinct; pale, spotless beneath; feet, 1. 4. 2. 3, first pair much the longest. A small species.

*Observations.* This makes a web like *Th. blandum.*
*Habitat.* Alabama. September.

## 20. Theridion sphærula.

Pl. 16, fig. 22.

*Description.* Yellow; cephalothorax with a black band, or
wholly black; abdomen subtriangular, orange, with a yellow
spot on the disk; one spot at each external angle, and region
of the nipples black; sometimes it is black or deep rufous
above and beneath, except the yellow spot on the disk, and two
little yellow dots near the base; feet pale yellow, 1. 2. 4. 3. A
very small species.

*Observations.* This very variable species is nevertheless
readily recognized by the shape of its abdomen. It is common,
and makes an orbicular white cocoon placed in its web.

*Habitat.* Alabama. May–September.

[♀ 2.8 mm. long; ♂ 2 mm. long.
Pl. 21, figs. 17, 17a, palpus of ♂.
Holyoke, Mass., July 4, ♂ and ♀; Peabody, Mass., June, ♂ and ♀;
Washington, D. C. E. P. Austin. J. H. E.].

## 21. Theridion globosum.

Pl. 16, fig. 23.

*Description.* Black; abdomen truncated behind, truncated
area whitish, with an obscure spot, and obsolete marks; feet
1. 4. 2. 3. A very small species.

*Observations.* This very distinct little Theridion was found
in its web on the stump of a tree. Its cocoons, quite numerous,
were of a pale cream color, tapering at both ends equally.
Young spiders were issuing from one of them. Specimens,
evidently of the same species, were found in June, [280]
which were yellowish where this is black; otherwise agreeing
with this in form and marking.

*Habitat.* Alabama. August.

[Pl. 19, fig. 125, cocoons. The spider was found under bark
or stones. *Supplement.*]

Tribe II.  PARTITÆ.  *External eyes apart.*

## 22.  Theridion trigonum.

Pl. 16, figs. 24, 25.

*Description.*  Pale brown or yellowish; lower middle eyes
borne on tubercles; abdomen triangular, with changeable
rufous lines, chiefly on the sides; male rufous, abdomen triangular, narrower behind; feet 1. 4. 2. 3, or 1. 2. 4. 3.  A small
species.

*Observations.*  This species, though varying much in color
and marking, is at once recognized by the form of its abdomen,
which, when viewed sideways, appears three-sided.  It makes
the usual web of Theridion, but sometimes it has an additional
web, like that of Linyphia.  It is found very common in
autumn, constantly in an inverted position in the middle of its
web.  Its cocoon is of a very singular shape, being orbicular
and sometimes ovoid, with a neck turned downward, like an
inverted gourd, and suspended by a thread attached to the web.
One of those cocoons being opened was found to contain the
pupa of an hymenopterous insect, a parasite.

*Habitat.*  Alabama.  July–September.

[Pl. 19, fig. 117, lateral view; fig. 131, cocoons.  Abdomen beneath variegated with rufous.  *Supplement.*]

[♀, length 4.2 mm.; cephalothorax 1.1 mm.; legs 7.2, 5, 3, 5.
♂, " 2.5 mm.; " 1.1 mm.; legs 8.2, 5.6, 3, 5.
Pl. 21, fig. 14, palpus of ♂.
The end of the abdomen is not pointed like fig. 25, pl. 16, but notched,
and in some specimens the two points are turned outward.  Specimens in
alcohol sometimes contrast like pl. 19, fig. 117.  The end of the abdomen
can be curved downward toward the spinnerets, and this is sometimes done
when the web is disturbed.
Beverly, Mass., May 10, on a birch twig, with an irregular web; June 18,
in the upper part of a web of *Linyphia scripta*; July 16, ♂ and ♀, among
the upper threads of a web of *Agelena nævia*; Holyoke, Mass., July 4, ♀, in
a tent made of one birch leaf hung in an irregular web; Peaks Island, Portland, Me., Aug. 11, cocoons of eggs on spruce trees.  J. H. E.]

### 23. Theridion verecundum.

Pl. 17, figs. 1, 2.

*Description.* Deep black, glossy; abdomen with blood-red spots underneath, which sometimes extend above in a chain, and with some waving white lines anteriorly, which are sometimes wanting: feet 1. 4. 2. 3. Male slender, abdomen with orange and white spots.

*Observations.* This spider, by its jet black color, is readily [281] distinguished. It is very common under stones, logs, or clods of earth, where it makes a web, the threads of which are so powerful as to arrest the largest hymenopterous insects, such as humble-bees. Its bite, if I can rely on the vague description of physicians unacquainted with entomology, is somewhat dangerous, producing alarming nervous disorders, which, however, are readily dispelled by brandy and other stimulants. There is no doubt that all spiders have a poison conveyed in the fang of their cheliceres, but in this case these organs are very small in proportion to the size of the spider, and, it would seem, are barely long enough to penetrate through the epidermis of a man's hand or foot.[1] The male, whose palpi have the black coil or penis external and very easily observed, is distinguished from the male of *Theridion lineatum* by that character, by its longer and slender legs, and by the white spots on the sides of the abdomen, which are not elongated in the form of lines. It has always been found on the top of weeds in a small web, and never under stones near the female. The cocoon is yellowish cinereous, of an ovoid form, and suspended by its pointed extremity.

*Habitat.* North and South Carolina, Georgia, Alabama, etc.

[It always remains in the centre of its web, feet uppermost. *Supplement.*]

[Ohio, ♀; Hog Island, Eastern coast of Virginia. Wm. Holden. J. H. E.]

---

[1] See Règne Animal, IV, 243, *A. mactans.*

### 24. Theridion lineatum.

Pl. 17, fig. 3.

*Description.* Cephalothorax blackish; abdomen deep purplish, or reddish black, with several diagonal white lines, and a
succession of red spots edged with yellow, and sometimes united
in the form of a band; a red spot underneath also; feet blackish, usually varied with yellow, 1. 4. 2. 3. Male with the
same markings.

*Observations.* This very common species is usually found
under stones, logs, or clods, always near the ground. It serves
as a prey to those singular hymenopterous insects, usually
called in the South *dirt daubers*, which enclose in [282] their
clay nests from twenty to forty small spiders, to serve as food
for their progeny. I once counted thirty-eight specimens of
this species extracted from one cell, made by a Trypoxylon, and
I have found them repeatedly in the nests of *Sphex cyanea.*
There may be some difficulty in distinguishing the male of this
species from the male of *Theridion verecundum*; the differences
are pointed out in the description of that species. The male of
this has never been observed with a compound palpus; the last
joint was merely greatly enlarged, as in the plate; but in some
specimens the enlargement was less remarkable. Can it be
that none of the very many specimens observed by me were
yet adult, and that the compound parts of the male organ
appear only at a certain period? It is possible that the plate
representing the male of this must be referred to *Th. verecundum.*

*Habitat.* North Carolina, Alabama.

[A specimen was found in March, with black legs and no
white bands on the abdomen; others were found in July. *Supplement.*]

[Marietta, O., ♀. Wm. Holden. j. h. e.]

Tribe III. Ventricosæ. *Abdomen caudate, subtriangular.*

### 25. Theridion? fictilium.

Pl. 17, fig. 4.

*Description.* Pale silvery on the abdomen above, yellowish underneath, with an abbreviated blackish band from the nipple-like projection, tapering towards the apex. Legs long and excessively slender, 1. 4. 2. 3.

*Observations.* This spider makes a web like Theridion, and remains motionless in an inverted position. The projection of the abdomen is capable of bending over nearly double. The markings of the male and female are alike. It is closely related to *T. intentum.*

*Habitat.* Alabama. July–September.

[♀, length 3.7 mm.; cephalothorax .8 mm.; legs 6, 3.4, 1.9, 4. Peabody, Mass., Sept. 28, near Bartholomew's pond. J. H. E.]

### 26. Theridion? pullulum.

Pl. 17, fig. 5.

*Description.* Animal yellow, with a longitudinal, forked, median brown line on the thorax; sides and the central line [283] of the back dark brown, and on each side of the latter four ocellated brown dots; legs long and slender, 1. 4. 2. 3.

*Observations.* This little spider makes a thin looking web, somewhat like Theridion, and dwells in dark places, in folds of paper, old rags, etc.

*Habitat.* Alabama.

[The spots on the abdomen are composed of little white dots surrounded by the brown marks. *Supplement.*]

### 27. Theridion? pertenue.

Pl. 17, fig. 6.

*Description.* Very small ; cephalothorax, abdomen and palpi black ; feet rufous, 4. 1. 2. 4.

*Observations.* Found usually under stones.
*Habitat.* Alabama.

### Genus Spintharus Mihi.

Characters. *Chelicres very slender; maxillæ slightly inclined towards the lip, widest at base, obliquely truncated above; lip short, wider at base, slightly emarginate; eyes eight, equal, disposed in the form of an ellipse open towards the base, two external eyes touching; feet long, slender, fourth pair longer than the first, the third being the shortest.*

*Habits.* Araneides sedentary, probably making an irregular web, composed of threads crossed in all directions, suspending themselves from a single thread, and thus capturing insects. Cocoon unknown.

*Remarks.* The species which serves as the type of this new subgenus could not be referred to Theridion. This will be obvious, when it is observed that its characters all depart from those of that natural subdivision, particularly its maxilla, which approaches the form of that organ in Epeira; by the position of its eyes, and the respective length of its feet, this spider would seem to approach wandering Araneides. It may ultimately be located among these. [284.]

### Spintharus flavidus.

Pl. 17, fig. 8.

*Description.* Yellowish; abdomen orange yellow, edge white, disk with a yellow margin, and spots surrounded and crossed in two places by a scarlet line, orange yellow, spotless beneath; feet, first and fourth pair with the antepenultimate joint tipped with orange, $\overline{4}$. 1. 2. 3.

*Observations.* This spider was found hanging by a thread from a tree thirty or forty feet high. It had secured, while thus suspended in the air, a wasp (Vespa), which, though many times larger than itself, was safely bound up for a repast.

*Habitat.* Alabama. September, October.

## Genus Pholcus Walck.

Characters. *Chelicercs small, cylindrical; maxillæ long, tapering to a point, inclined over the lip; lip widest near the base, short; eyes eight, subequal, two in the middle in a transverse row, three on each side placed together in the form of a triangle; feet excessively long, first pair longest, then the second, the third being the shortest.*

*Habits.* Aranecides sedentary, making in dark corners a very loose web of slender threads, crossed in all directions. Eggs collected together without a silk covering, which the mother carries with her cheliceres.

*Remarks.* This subgenus, by the extreme length of its legs resembles Phalangium. The species belonging to it may be found in apartments seldom visited, particularly churches or caves. They shake their body, when threatened by an enemy; but seem to have very weak means of offence, and to feed on the very smallest prey.

### Pholcus atlanticus.

Pl. 17, fig. 7.

*Description.* Pale or livid yellow; abdomen with more or less distinct lines and spots; cheliceres articulated together [285] near the middle; sometimes attaining four inches from the end of the anterior to that of the posterior leg.

*Observations.* This apparently powerless spider, no doubt related to the *Aranea phalangioides* of Europe, is found in the dark corners of the ceilings of uninhabited houses, in loose webs scarcely strong enough to detain any, even small insects. It is inactive, and never was seen by me with any prey, or with the show of obtaining any. This ought not to be mistaken for the Phalangium which children call daddy-long-legs. The female carries her eggs glued together, without a cocoon, in her cheliceres.

*Habitat.* Southern States. Alabama, at the entrance of limestone caves.

A female was found in Alabama in June, with a body resembling an orbicular cocoon, which she carried in her cheliceres. On tearing the silk covering, it was found to contain a Clubiona, which was thus wrapped up for future meals.

[♀, length 8 mm.; cephalothorax 2 mm.; legs 48.5, 35, 27.2, 34.5.
♂, " 6.4 mm.; " 2 mm.; legs 47.5, 34.5, 25.4, 30.
Pl. 21, fig. 18, palpus of ♂.
Salem, Mass., July 10, in a cellar near a window. When disturbed they hung by their feet and swung the body round in a circle so fast as to be hardly visible. July 12, females carrying cocoons of eggs in their mandibles. The cocoons were so thin that eggs could be plainly seen. Eggs laid July 30, hatched Aug. 11.
*Pholcus phalangioides* Blackw., Spiders of Gt. Britain and Ireland. J. H. E.]

## Subgenus OOPHORA[1] Mihi.

[*Spermophora* Hentz] Silliman's Journal, Vol. XLI, p. 116.

*Characters. Cheliceres short, cylindrical; maxillæ wide at base, tapering to a point, inclined over the lip; lip short, widest at base; eyes six, equal, three on each side, placed together in the form of a triangle; feet slender, moderately long, first pair longest, the fourth and the second nearly equal, the third shortest.*

*Habits.* Araneides sedentary, making in obscure places an excessively loose and slender web, composed of a few threads crossed in various directions. Eggs not enclosed in a cocoon, but agglutinated together, which the mother carries between her cheliceres.

*Remarks.* This subgenus is very closely related to Pholcus. Nay, had it eight eyes instead of six, and were its legs much longer, it could not have been separated from that subdivision [286] of Aranea. But it is obvious that these characters require the separation.

The spider upon which the subgenus is constituted has habits

[1] [Originally published as *Spermophora*. See "Silliman's Journal," as above cited; also this work, p. 14, where the article from Silliman's Journal is reprinted.—ED.]

similar to those of Pholcus. It does not dwell in walls, but seeks dark nooks under any kind of rubbish which has been long neglected, and, when disturbed, runs off with its progeny, if it have any, and seeks for some darker place undisturbed by the broom of the housewife. It must live on microscopic animalcules, owing to its diminutive size, and the weakness of its threads.

### Oophora meridionalis.

[ *Spermophora meridionalis* Hentz] Silliman's Journal, Vol. XLI, p. 116. Pl. 17, fig. 9.

*Description.* Livid white or pale yellow above and beneath; cephalothorax with two small, angular, plumbeous spots.

*Observations.* This small spider is common in dark corners and obscure apartments, where it makes loose, slender webs, in the manner of Pholcus. The female is always found with her eggs, when she has them, carrying them in her cheliceres. These eggs are not enclosed in a cocoon, but glued together in a mass consisting of from ten to fifteen.

*Habitat.* North Alabama.

[?, length 2 mm.; cephalothorax .6 mm.; legs 8.7, 6.4, 5.2, 6.8. Salem, Mass., Apr. 13, young in a closet; Boston, Mass., June 8, in a drawer in the Museum. (Marietta, Ohio. Wm. Holden.) J. H. E.]

### Mygale fluviatilis.

Pl. 17, fig. 15.

*Description.* Livid; cephalothorax with a depression near the middle above; abdomen with two transverse lines near the base; third pair of legs sensibly larger, though shorter than the rest. Feet 4. 1. 2. 3.

*Observations.* This new species was found in the water during an inundation of the Tennessee River. It has been found since in its hole, deep in the ground. The tubular cavity, at least a foot in depth, was supplied with a *door* or silken lid closing the aperture.

*Habitat.* Alabama. March, October. [287.]

## Subgenus KATADYSAS Mihi.

Characters. *Eyes eight, subequal, in two rows, both curved downwards; fang of the cheliceres articulated downwards; palpi inserted near the extremity of the maxillæ. Feet 4. 1. 2. 3. Pulmonary orifices only two.*

Observations. This very anomalous spider, found only once, offers a very striking instance of the manner in which nature combines characters, so as to separate widely animals which are apparently closely allied. This has all the essential characters of Mygale, but one (having but two pulmonary orifices), and yet it is obviously related to Lycosa, near which it should be placed in a natural arrangement. I know nothing of its habits, except that it dwells or hides under stones. It probably makes no web.

*Habitat.* Alabama.

### Katadysas pumilus.

Pl. 17, fig. 16.

*Description.* Livid, testaceous; cephalothorax with two longitudinal bands near the middle, and two curved fillets near the edge, fuscous; abdomen with a line bifurcated anteriorly on the middle, and two lines of minute dots on the sides, fuscous; same color underneath, with minute fuscous dots.

*Habitat.* North Alabama. Under stones.

### Micrommata pinicola.

Pl. 17, fig. 14.

*Description.* Whitish, cephalothorax with the area of the eyes dusky; abdomen with various indistinct curved lines and impressions; venter with two curved lines of minute brown dots. Feet 1. 2. 4. 3.

*Habitat.* South Alabama. [288.]

### Micrommata subinflata.

Pl. 17, fig. 13.

*Description.* Livid testaceous; cephalothorax with a dusky longitudinal band; abdomen with angular dusky spots forming a longitudinal band; feet tipped with dusky. Feet 2. $\overline{4.1}$, or $\overline{1.}$ 4. 3.

*Observations.* This and the preceding species have considerable affinity with Dolomedes. In fact, it is difficult to trace the exact limits between the two subgenera.

*Habitat.* South Alabama. In dark places, on the ground.

### Attus sinister.

Pl. 17, fig. 12.

*Description.* Black, varied with rufous; abdomen whitish at base; venter with an interrupted ash-colored band; feet 4. $\overline{1.}$ 2. 3.

*Observation.* This spider should be placed in my tribe of the Luctatoriæ.

*Habitat.* Alabama.

### Attus retiarius.

Pl. 17, fig. 11.

*Description.* Livid greenish; cephalothorax with an indistinct brown spot; abdomen with two abbreviated brownish bands, approaching towards the apex.

*Observations.* This Attus was discovered and delineated by my son, Charles A. Hentz, whose attention is more particularly drawn towards the study of Ichthyology. He found the female devouring her male. I believe the markings of the male differ from those of the female. It belongs to my tribe of the Metatoriæ.

### Synemosyna noxiosa.

Pl. 17, fig. 10.

*Description.* Piceous; abdomen very slightly contracted

near the base, with an interrupted whitish line across.  Feet
1. 4. 2. 3. ; first pair stout.
*Habitat.*  Alabama.

---

[From the Journ. Acad. Nat. Sciences, II, 53.]

A NOTICE CONCERNING THE SPIDER WHOSE WEB IS USED IN
MEDICINE.  BY N. M. HENTZ.

It has been found lately, that the web of a species of spider,
common in the cellars of this country, possesses very narcotic
powers, and it has been administered apparently with success in
some cases of fevers.

Having for some time past, studied with care the genus Aranea of Linneus, I have been induced to write a description of
this species; I therefore made a drawing taken from a large
female, which accompanies the present notice.

The genus Aranea of the first writers on Entomology being
composed of a very great number of species, it has been found
necessary to divide it into smaller sections, or families.  Gmelin's
edition of Linneus contains ninety-eight species; Walckenaer
enumerates nearly three hundred, and the number may be carried to a thousand.  If the colour of the abdomen were the
only character to find the species among several hundreds, it
would be a very difficult task to assign with certainty a name to
each separately, without any other description.  Messrs. Latreille and Walckenaer have rendered the history of this genus
quite easy to study: they have left little undone in regard to
the species known to them.  It is to be regretted that Mr.
Walckenaer's Tableau des Aranéides is not a more common
work.

I shall therefore give the generic characters of this spider, as
if the work were unknown to the naturalists in this country.

It belongs to the genus Tegeneria of Walckenaer, and to
that of spiders, properly so called, of Latreille.  Its characters

are : eight eyes, forming two parallel lines, the upper being curved and longer. Lip wider in the middle, cut straight at its extremity. Maxillæ inserted upright, not bent on the lip. Corselet nearly as large as the abdomen. The first pair of legs the longest, the fourth next, then the second, and the third the shortest.

*Manners.* Spiders forming an horizontal web, with a cylindrical tube, in the form of a funnel.

This is sufficient to characterise the genus, containing the different species of spiders which inhabit cellars and dark places. The species that makes its web in the fields, on bushes, does not belong to the same genus; it has been properly separated from it by Walckenaer. The last pair of legs is the longest in this, and the eyes differ essentially in their situation. There is another species, very common in Carolina, which, however, I have not yet observed here, making a web nearly similar to this, but very different in all its generic characters; it ought not to be taken for the other : I intend publishing a description of the genus Aranea, in which this will form a separate section. But the characters which I have given are sufficient to ascertain whether a spider belongs to the genus Tegeneria, so that with some attention, no mistake will occur.

The species which I am treating of, is of a black colour, inclining to blue; the abdomen is marked with about ten livid pale spots, and a line toward its anterior extremity : I have seen specimens where the legs were marked with black spots. I think it necessary to remark here, that spiders of the same species living in dark places, vary greatly in their colours, according to the manner in which the light strikes upon them. The great point in this case I think, is to ascertain the genus, for it appears that the web of all species belonging to it has the same virtues, and this is distinct from the *Aranea domestica,* whose web has been used in Europe : we see an illustration of this in the genus Meloë, where every species possesses more or less the blistering power.

The present American spider, I think, has not been as yet described: for the present I shall call it *Tegeneria medicinalis*. —Pl. V. fig. 1.[1]

*a* — organs of manducation.

*b* — position of the eyes.

[1] [This figure being very poor has not been reproduced, the species is moreover figured on Plate 11, (f. 21.). See also p. 99. Ed.].

# INDEX.

(165)

## PLATE I.

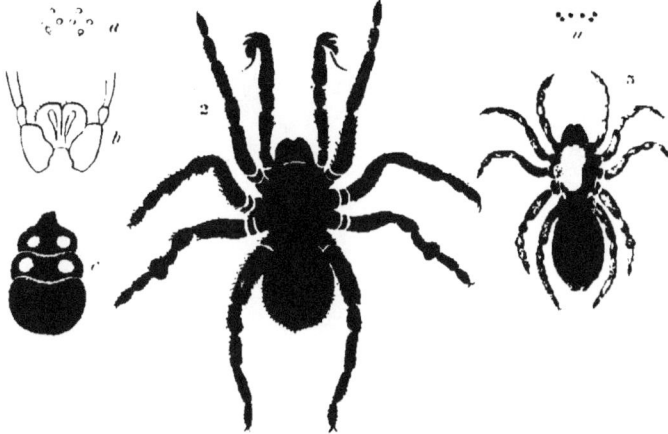

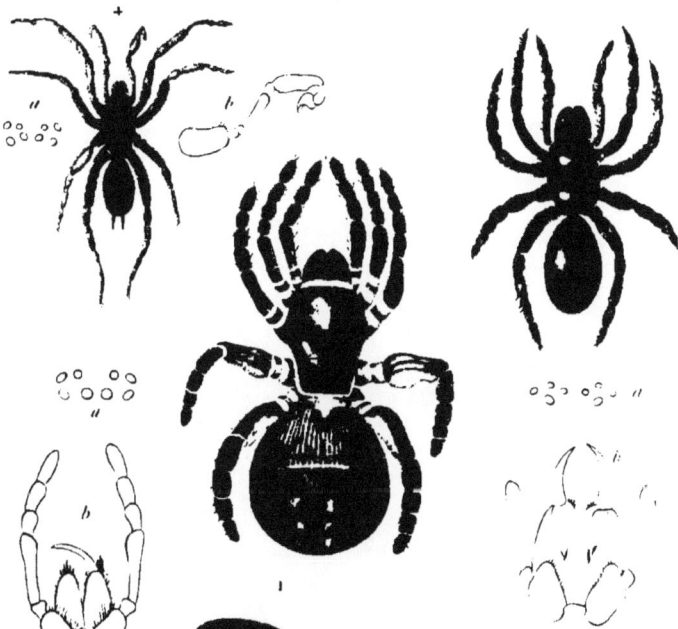

1 MYGALE *truncata*
2 *solstitialis*  3 M. *Carchmensis*  4 M *gracilis*
*// M. (gregg) J. ft.*  5 *unicolor*

# PLATE II.

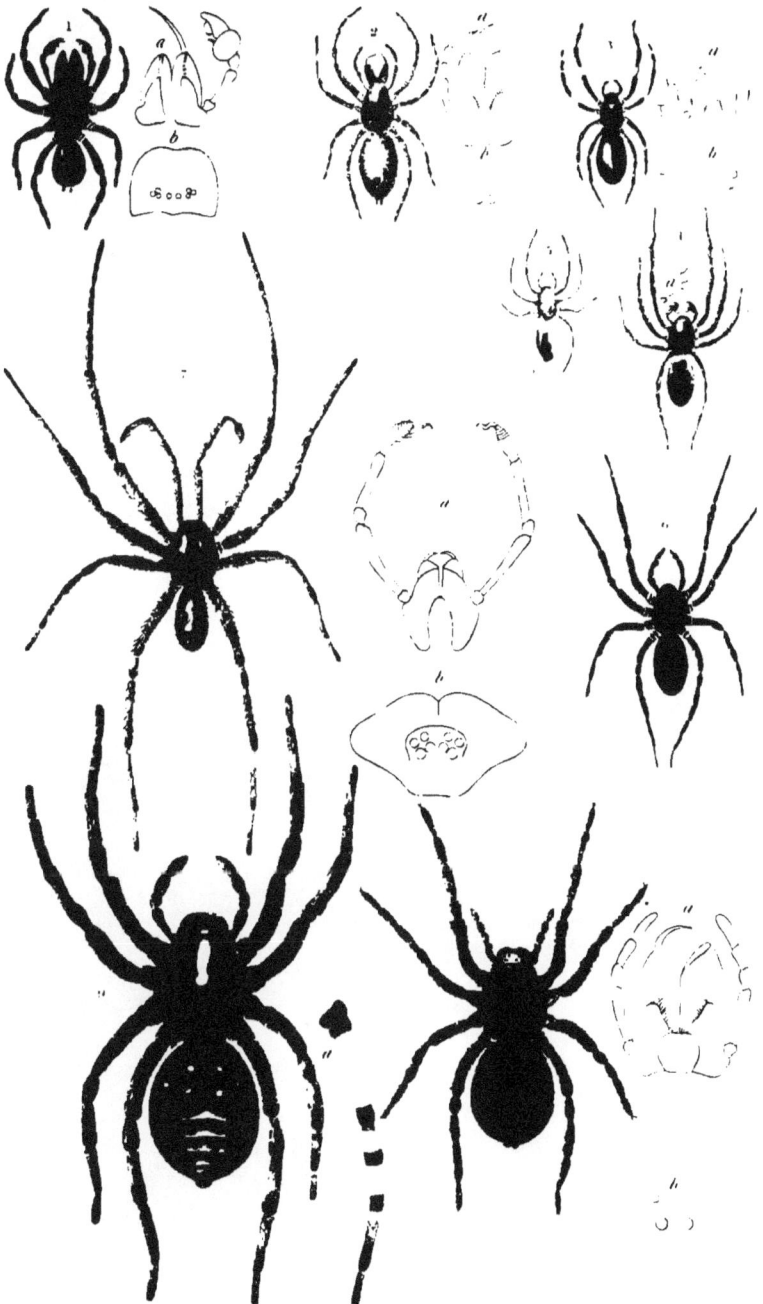

# PLATE III.

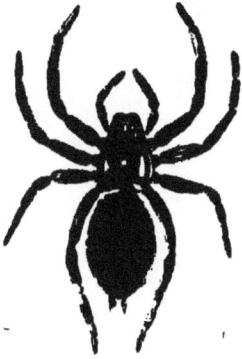

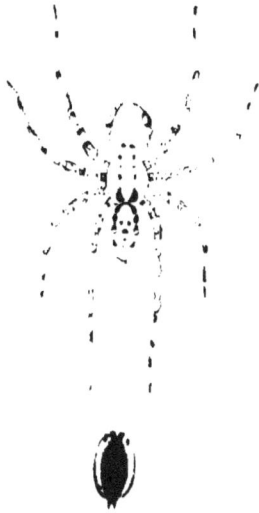

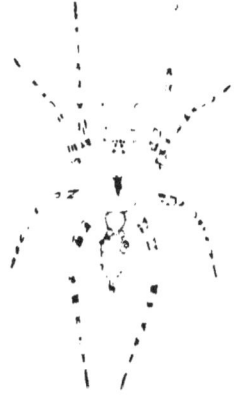

# PLATE IV.

# PLATE V.

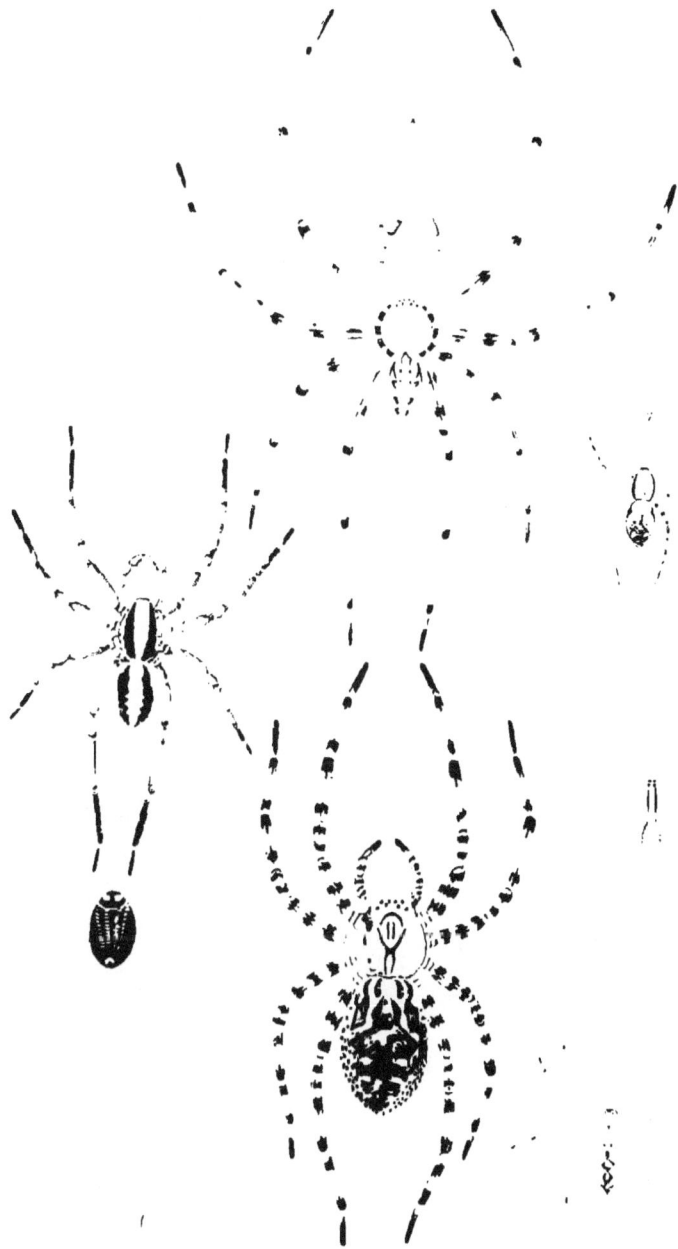

# PLATE VI.

Fig. 1. Dolomedes scriptus.
" 2.  "   albineus.
" 3.  "   urinator.
" 5.  "   sexpunctatus, ♂, young.
" 6.  "    "   ♀.
" 7. Micrommata undata.
" 8.  "   serrata.
" 9.  "   carolinensis, young.
" 10. Oxyopes salticus.

PLATE VII.

# PLATE VIII.

Fig 1

Fig 2

Fig 3

Fig 4

Fig 5

Fig 6

Fig 7

Fig 8

Fig 12

Fig 9

Fig 10

Fig 11

Fig 15

Fig 12

Fig 14

Fig 13

Fig 16

Fig 17

Fig 18

Fig 19

Fig 20

Fig 21

Fig 22

A. Weber del.

V.H. Tappan sc.

Hentz's Araneides of the U. States

# PLATE IX.

VOL. V      PL. XXII

Fig 24

Fig 23

Fig 1

Fig 2

Fig 3

Fig 4

Fig 5

Fig 6

Fig 7

Fig 8

Fig 9

Fig 10

Fig 11

Fig 12

Fig 13

Fig 14

Fig 15

Fig 17

Fig 16

Fig 18

Fig 19

Fig 20

Fig 21

Hentz's Araneides of the U. States

## PLATE X.

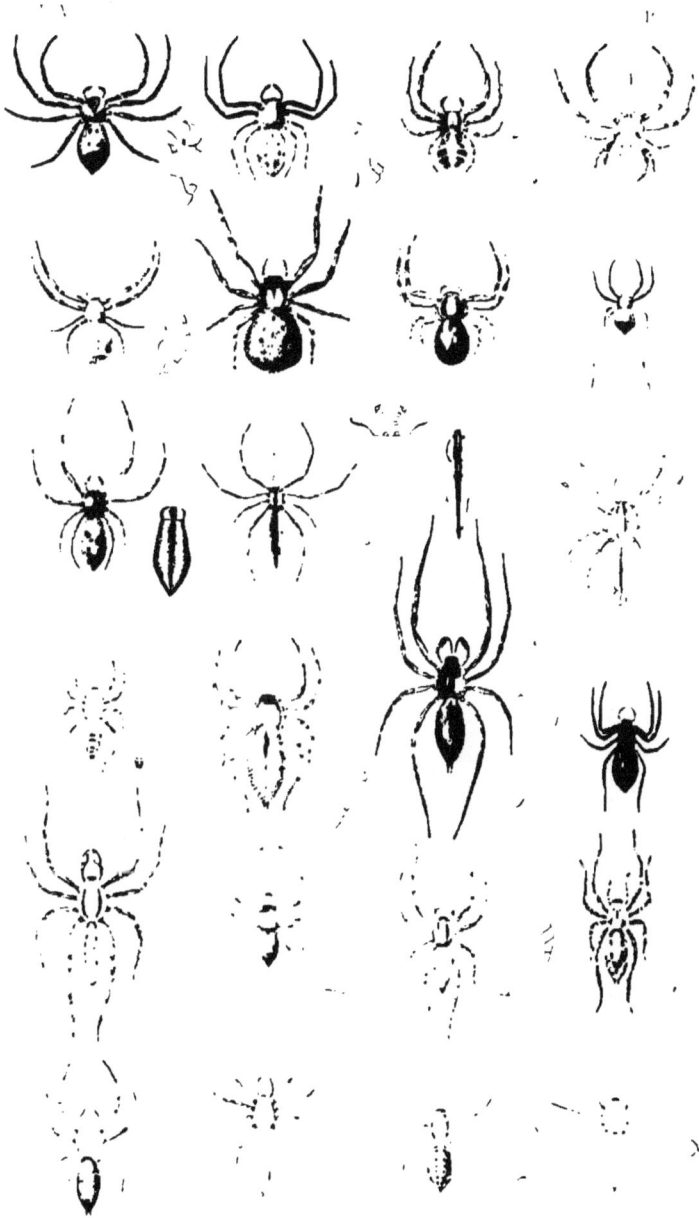

# PLATE XI.

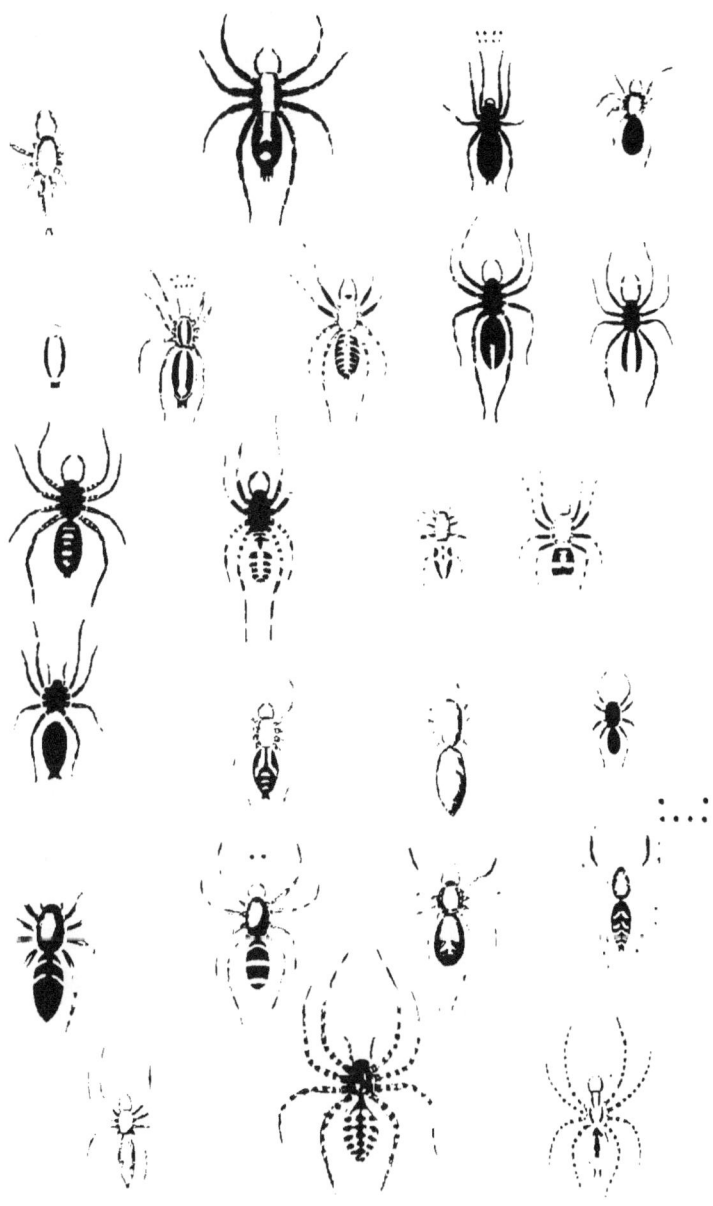

# PLATE XII.

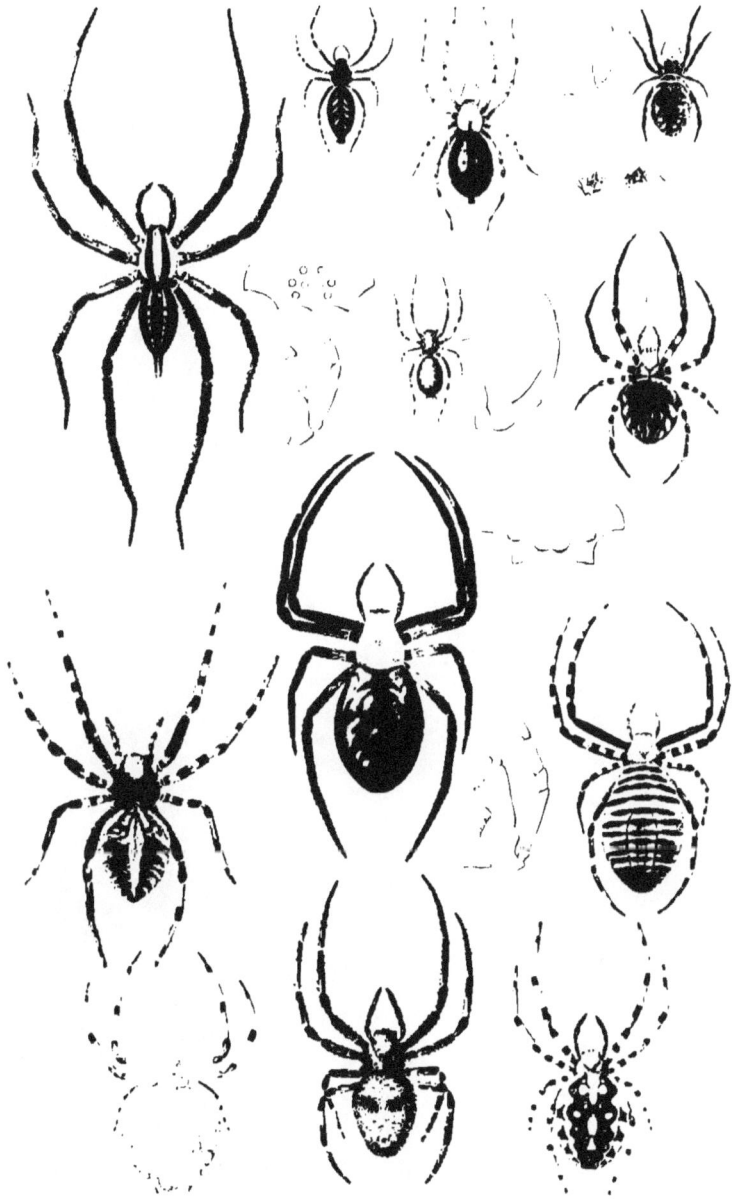

# PLATE XIII.

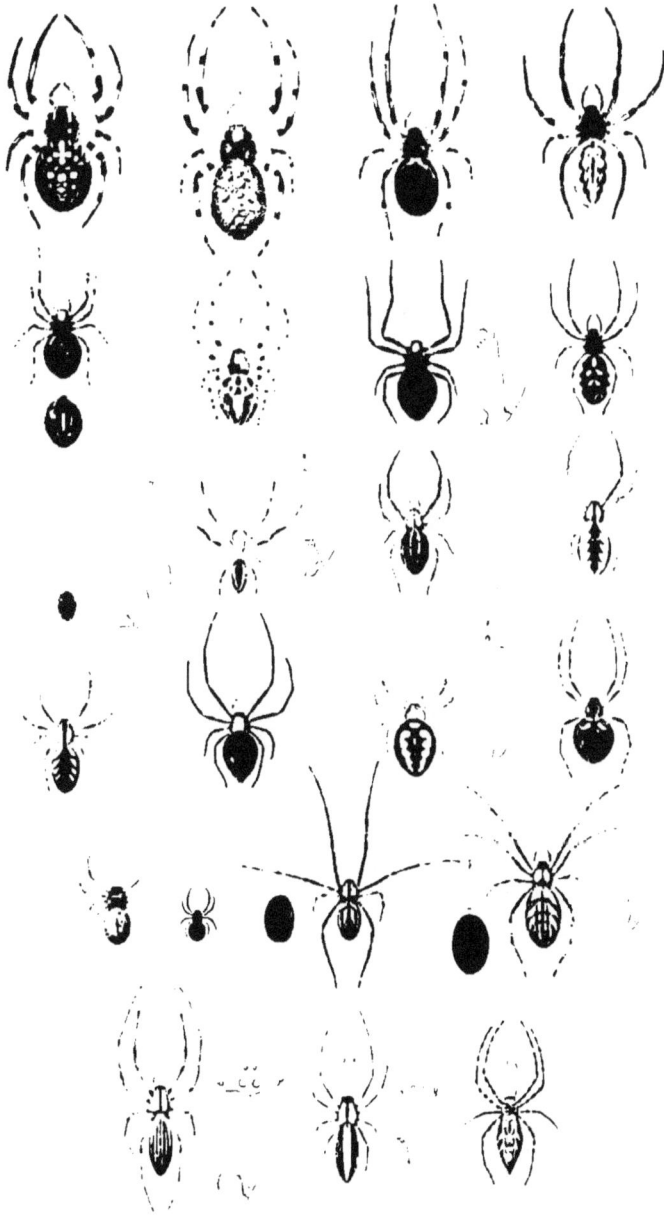

## PLATE XIV.

# PLATE XV.

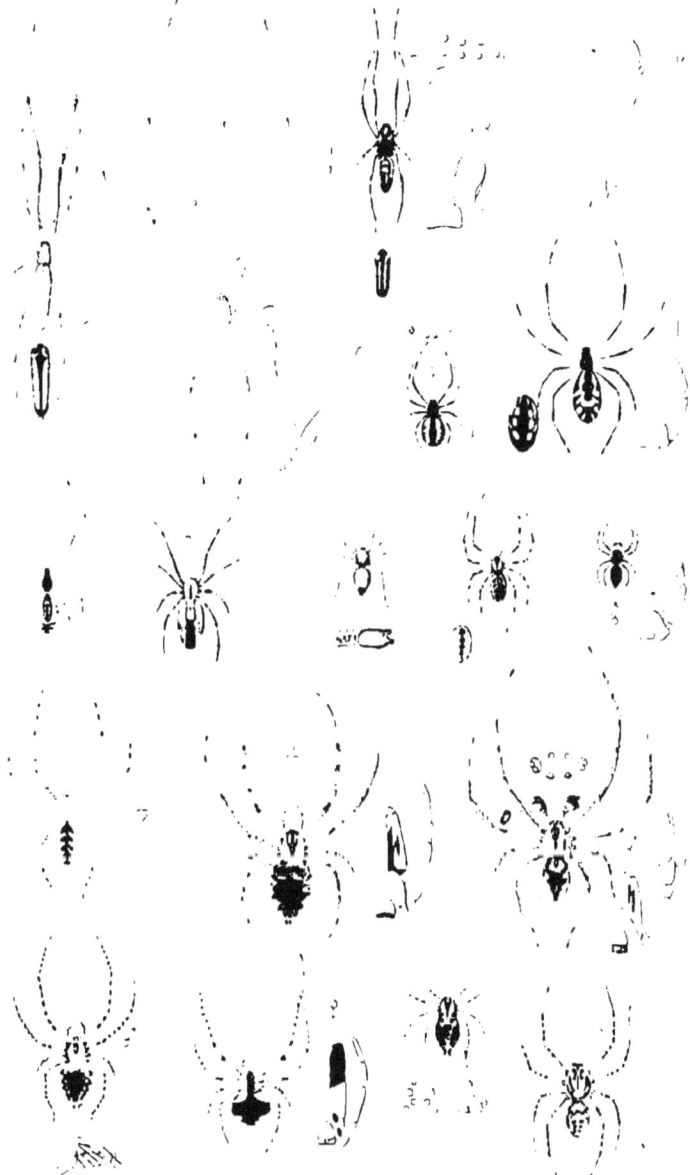

# PLATE XVI.

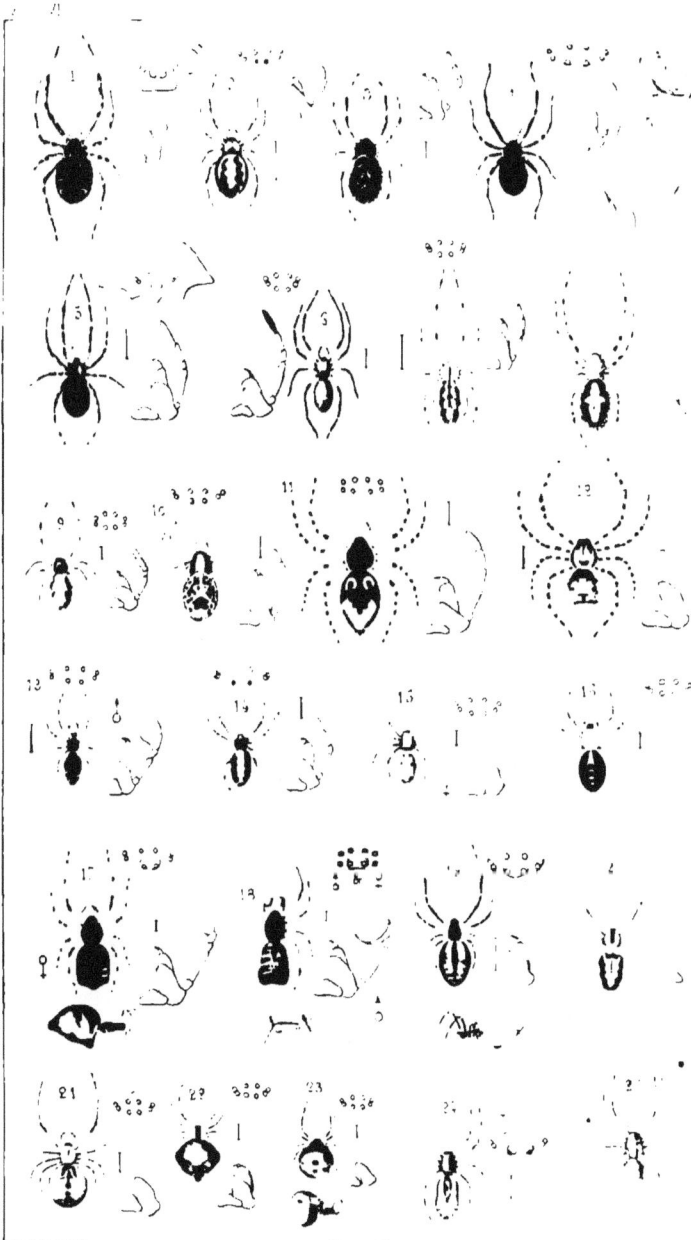

## PLATE XVII.

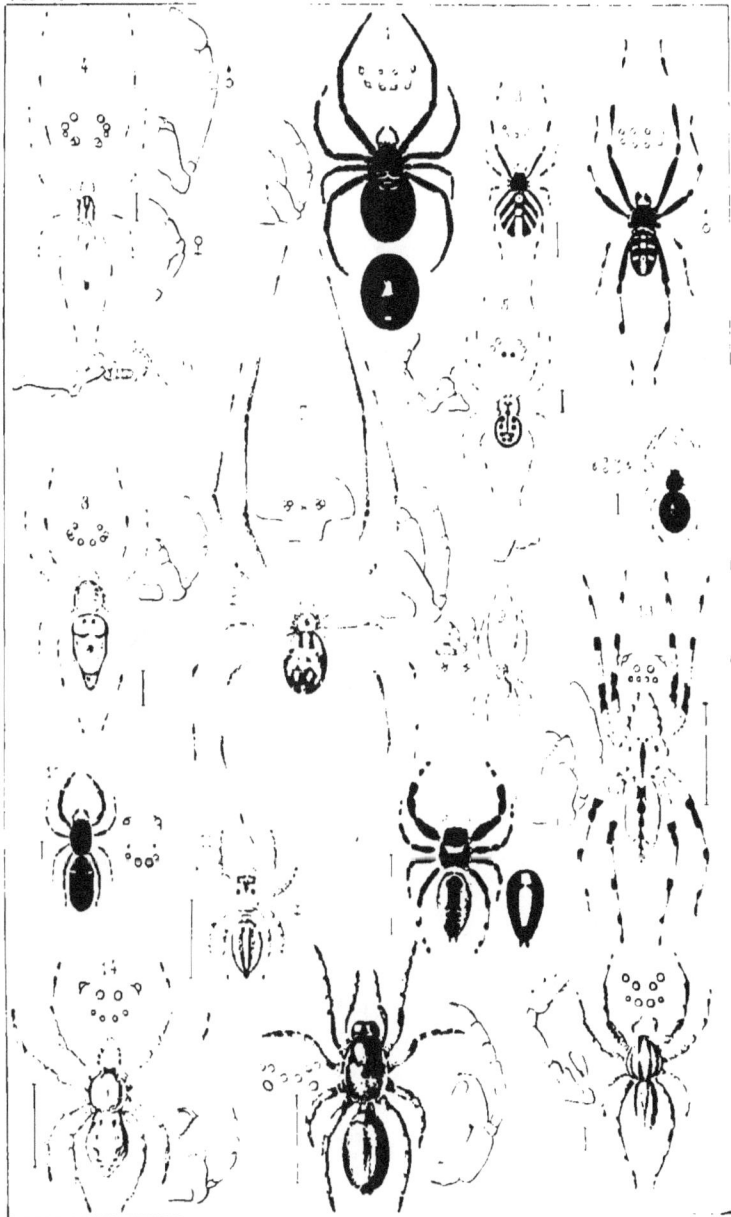

PLATE XVIII.

Fig. 1. Micrommata serrata.
" 2. Attus elegans.
" 3. " hebes.
' 4. " niger.
5. " superciliosus.
6. " canonicus.
" 7. Clubiona pallens.
" 8. " gracilis.
' 9. Prodidomus rufus.
" 10. Epeira hortorum.
" 11. " gibberosa.
' 12. Linyphia coccinea.
" 13. " neophyta.
" 14. Oxyopes viridans.
' 15. Attus roseus.
" 16. Clubiona obesa.
" 17. " pallens.
" 18. " celer.
" 19. " saltabunda.
" 20. Epeira pentagona.
" 21. " alba.
" 22. " mitrata.
" 23. Linyphia marmorata.
" 24. " costata.
" 25. Epeira labyrinthea.
" 26. Attus capitatus.
" 27. " nubilus.
" 28. " puerperus.
" 29. Clubiona piscatoria.
" 30. Epeira placida.
" 31. " spiculata.
" 32. Clubiona albens.
" 33. Minnetus interfector.
" 34. " syllepsicus.
" 35. Attus fulcarius.
' 36. " castaneus.
" 37. " rufus.
" 38. " sinister.
" 39. Thomisus aleatorius.
" 40. " piger.
" 41. " asperatus.
" 42. " parvulus.
" 43. Clubiona agrestis.
" 44. Epeira cornigera.
" 45. Agelena plumbea.
" 46. Epeira obesa.
" 47. " prompta.
" 48. " nivea.
" 49. " hamata.
" 50. " foliata.
" 51. " displicata.
" 52. " infumata.
" 53. " heptagon.
" 54. " caudata.
" 55. Dolomedes sexpunctatus.
" 56. Micrommata marmorata.

Fig. 57. Attus gracilis.
" 58. " sylvanus.
" 59. Epiblemum faustum.
" 60. Thomisus caudatus.
" 61. Phillyra riparia.
" 62. Epeira sanguinalis.
" 63. Attus fasciolatus.
" 64. " viridipes.
" 65. " auratus.
" 66. " cyaneus.
" 67. Synemosyna scorpiona.
" 68. " ephippiata.
" 69. Thomisus fartus.
" 70. Epeira sutrix.
" 71. Linyphia autumnalis.
" 72. Epeira heptagon.
" 73. Dolomedes albineus.
" 74. Attus familiaris.
" 75. " tripunctatus.
" 76. " mystaceus.
" 77. Thomisus vulgaris.
" 78. " celer.
" 79. " Duttonii.
" 80. Cyllopodia cavata.
" 81. Epeira hebes.
" 82. Attus coronatus.
" 83. Thomisus ferox.
" 84. " tenuis.
" 85. Clubiona tranquilla.
" 86. " inclusa.
" 87. " immatura.
" 88. Epeira trifolium.
" 89. " stellata.
" 90. Oxyopes salticus.
" 91. Lyssomanes viridis.
" 92. Attus auratus.
" 93. Epeira labyrinthea.
" 94. " spiculata.
" 95. " sutrix.
" 96. " caudata.
" 97. Linyphia autumnalis.
" 98. Micrommata undata.
" 99. Attus familiaris.
" 100. Thomisus caudatus.
" 101. " tenuis.
" 102. Clubiona tranquilla.
" 103. Epeira trifolium.
" 104. Linyphia communis.
" 105. Micrommata marmorata.
" 106. Attus tripunctatus.
" 107. " gracilis.
" 108. " sylvanus.
" 109. Epiblemum faustum.
" 110. Tegenaria medicinalis.
" 111. Phillyra riparia.

# PLATE XIX.

Fig. 112. Attus cristatus.
" 113. Theridion subulatum.
" 114. Synemosyna ephippiata.
" 115. Linyphia conferta.
" 116. Epeira caudata.
" 117. Theridion trigonum.
" 118. Linyphia communis.
" 119. Attus mystaceus.
" 120. Oxyopes scalaris.
" 121. Epeira riparia.
" 122. " rugosa.
" 123. " domiciliorum.
" 124. " labyrinthea.
" 125. Theridion globosum.
" 126. Phyllyra mammeata.
" 127. Mimetus interfector.
" 128. Theridion morologum.
" 129. " roscidum.
" 130. Linyphia scripta.
" 131. Theridion trigonum.
" 132. Epeira caudata.
" 133. " labyrinthea.
" 134. Oxyopes viridans.

PLATE XX.

Fig. 1. Dysdera interrita, palpus of ♂.
" 2. Pylarus bicolor, foot.
" 3. Lycosa carolinensis, palpus of ♂.
" 4. Micrommata carolinensis, ♂, ♀, and palpus of ♂.
" 5. Attus familiaris, palpus of ♂.
" 6. " tripunctatus, palpus of ♂.
" 7. " mystaceus, palpus of ♂.
" 8. Epiblemum faustum, ♂, ♀, and palpus of ♂.
" 9. Synemosyna fumica, palpus of ♂.
" 10. Thomisus vulgaris, palpus of ♂.
" 11. " Duttoni, palpus of ♂.
" 12. Clubiona obesa, palpus of ♂.
" 13. " pallens, palpus of ♂.
" 13a. " " " "
" 14. Herpyllus alarius, palpus of ♂.
" 15. Clubiona saltabunda, palpus of ♂.
" 16. Herpyllus ater, palpus of ♂.
" 17. " variegatus, palpus of ♂.
" 17a. " " " "
" 18. " descriptus, palpus of ♂.
" 19. Tegenaria medicinalis, palpus of ♂.
" 20. Agelena nævia, palpus of ♂.
" 21. Cyllopodia cavata, ♂, ♀, palpus of ♂, and calamistrum.

# PLATE XXI.

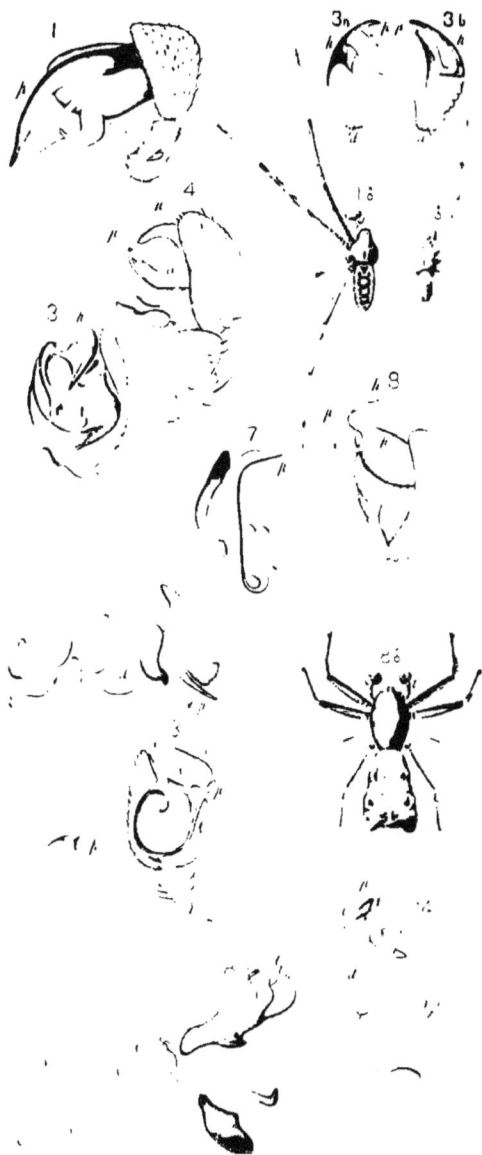